# Springer Series in Solid-State Sciences

## Volume 193

**Series editors**

Bernhard Keimer, Stuttgart, Germany
Roberto Merlin, Ann Arbor, MI, USA
Hans-Joachim Queisser, Stuttgart, Germany
Klaus von Klitzing, Stuttgart, Germany

The Springer Series in Solid-State Sciences consists of fundamental scientific books prepared by leading researchers in the field. They strive to communicate, in a systematic and comprehensive way, the basic principles as well as new developments in theoretical and experimental solid-state physics.

More information about this series at http://www.springer.com/series/682

Shashikant Mulay · John J. Quinn
Mark Shattuck

# Strong Fermion Interactions in Fractional Quantum Hall States

## Correlation Functions

Shashikant Mulay
Department of Mathematics
University of Tennessee
Knoxville, TN, USA

John J. Quinn (*deceased*)
Knoxville, USA

Mark Shattuck
Department of Mathematics
University of Tennessee
Knoxville, TN, USA

and

Researcher at the Institute for Computational
    Science (2017–present)
Ton Duc Thang University
Ho Chi Minh City, Vietnam

ISSN 0171-1873                ISSN 2197-4179   (electronic)
Springer Series in Solid-State Sciences
ISBN 978-3-030-13118-0        ISBN 978-3-030-00494-1   (eBook)
https://doi.org/10.1007/978-3-030-00494-1

This Springer imprint is published by the registered company Springer Nature Switzerland AG
The registered company address is: Gewerbestrasse 11, 6330 Cham, Switzerland

# Preface

A trial wave function $\Psi(z_1, \ldots, z_N)$ of a system of Fermions can always be expressed as the product of an antisymmetric Fermion factor $\prod_{1 \le i < j \le N}(z_i - z_j)$, and a symmetric correlation factor $G := G(z_1, \ldots, z_N)$ arising from Coulomb interactions. Here, $z_i$ is the complex coordinate of the $i$th Fermion. One can represent the Coulomb interactions diagrammatically as a multigraph on $N$ vertices with edges representing the correlation factors. For Jain states at filling factor $\nu = p/q < 1/2$, the value of the single particle angular momentum $l$ satisfies the relation $2l = \nu^{-1}N - c_\nu$, where $c_\nu = q + 1 - p$ is the finite size shift. The value of $(2l, N)$ defines the function space of the $2l + 1$ states into which one must insert $N$ Fermions. This imposes a number of conditions on the correlation factor $G$. Knowing the value of the total angular momentum $L$ for IQL states and for states containing quasielectrons (or quasiholes) from Jain's mean field composite Fermion picture allows one to determine the exact conditions $G$ must satisfy. The dependence of the pair interaction energy $V(L_2)$ on the pair angular momentum $L_2$ suggests a small number of correlation diagrams for a given value of $L$. We have proposed an intuitive approach which determines the symmetric correlation factor $G$ associated with a correlation diagram. In [1–4], this intuitive approach and its applications are briefly presented. For small values of $N$, our approach is justified via the observed agreement with numerical diagonalization studies. For systems in Jain IQL states, our approach has led to the discovery of 'minimal (energy) configurations' corresponding to all filling factor $\nu < 1/2$. These minimal configurations are easily appreciated as natural generalizations of the Laughlin configuration to the case of a general $\nu < 1/2$. The main objective of this monograph is to explain in detail the physics as well as the mathematics of our theory of correlation factors. Thus, Chap. 1 builds up a careful justification and detailed motivation for our intuitive approach, while the second builds the mathematical concepts and tools needed for rigorous proofs. The last two sections of Chap. 2 deal with concrete applications that focus on the computation of the correlation factors $G$. The classical theory of semi-invariants of binary forms is closely related to the mathematical considerations in our approach. Moreover, the theorems presented in the third section of Chap. 2 make a novel contribution to invariant theory, and hence, they are of interest from a purely mathematical standpoint. Our desire to present this attractive symbiosis between the theory of invariants and the correlation polynomials $G$ serves as one of the motivations for this monograph. The first Appendix A is devoted to proving that our intuitive approach is indeed applicable to the Moore-Read state and essentially reproduces the known trial wave function. In the second Appendix B, we pose some currently unresolved problems closely related to the mathematics in Chap. 2 and brought to light by our investigations [3, 4]. The last appendix records some light computational procedures that we have used. We have striven to make this monograph as broadly accessible as possible so that it can be used as a text in an advanced course dealing with the fractional quantum Hall effect.

Knoxville, USA                                                                    Shashikant Mulay
Knoxville, USA                                                                        John J. Quinn
Knoxville, USA/Ho Chi Minh City, Vietnam                                Mark Shattuck

# References

1. S.B. Mulay, J.J. Quinn, M.A. Shattuck, Correlation diagrams: an intuitive approach to correlations in quantum Hall systems. J. Phys: Conf. Ser. **702**, 1–10 (2016)
2. S.B. Mulay, J.J. Quinn, M.A. Shattuck, A generalized polynomial identity arising from quantum mechanics. Appl. Appl. Math. **11**, 576–584 (2016)
3. S.B. Mulay, J.J. Quinn, M.A. Shattuck, An algebraic approach to electron interactions in quantum Hall systems. arXiv:1809.01504 (2018)
4. S.B. Mulay, J.J. Quinn, M.A. Shattuck, An algebraic approach to FQHE variational wave functions. arXiv:1808.10284 (2018)

# Acknowledgements

First and foremost, we are very grateful to Luis Finotti for his frequent and generous aid in carrying out computations via SAGE as well as MAPLE; his expert help was essential in computing the third row of the last column of Table 2.2. We thank George Simion for doing some numerical testing for us. We are grateful to Brendan McKay for his advice on the *gtools* software [1] which allowed us to compute the entries of the second row of Table 2.2. Most of the correlation diagrams of this monograph were drawn using the *multigraph* package [2] of the R library; we are grateful to Antonio Rivero Ostoic for answering our queries about this package. We thank Abdelmalek Abdesselam and Jaydeep Chipalkatti for providing the reference [3]. Lastly, we thank Marie Jameson for providing a MAPLE procedure which lists the sets $E(N, d)$ (for arbitrary $N$ and $d$).

Knoxville, USA                                          Shashikant Mulay
Knoxville, USA                                          John J. Quinn
Knoxville, USA/Ho Chi Minh City, Vietnam               Mark Shattuck

# References

1. B.D. McKay, A. Piperno, Practical graph isomorphism, {II}. J. Symbolic Comput. **60**(0), 94–112 (2014)
2. A.R. Ostoic, *multigraph: plot and manipulate multigraphs* (2017). R package version 0.75
3. J. Dixmier, Quelques résultats et conjectures concernant les séries de Poincaré des invariants des formes binaires, in *Séminaire d'algébre Paul Dubreil et Marie-Paule Malliavin*, 36éme année (Paris, 1983–1984), Lect. Notes Math., vol. 1146 (Springer, Berlin, 1985), pp. 127–160

# Contents

1    **Fermion Correlations** ...............................    1
    1.1    Introduction ........................................    1
    1.2    The Integral and the Fractional Quantum Hall Effects .........    3
    1.3    Jain's Composite Fermion Approach .....................    4
    1.4    The Composite Fermion Hierarchy .....................    6
    1.5    Justification of the CF Approach ......................    9
    1.6    Numerical Diagonalization Studies ....................    13
    1.7    Correlations and Correlation Diagrams..................    14
    1.8    Correlation Diagrams for $N = 4$ .....................    18
    1.9    Extension to Larger Systems .........................    21
    References ............................................    23

2    **Correlation Functions** ...............................    27
    2.1    Introduction ........................................    27
    2.2    Invariant-Theoretic Essentials .......................    36
    2.3    Constructions of Semi-invariants .....................    56
    2.4    Fermions in the $\nu < 1/2$ IQL State..................    99
       2.4.1    $\nu = n/(2pn + 1)$ ...........................    100
       2.4.2    $\nu = n/(2pn - 1)$ ...........................    106
    2.5    Systems with QE in the $\nu = 1/3$ IQL ...............    111
    References ............................................    134

**Appendix A: Moore-Read State** .............................    137

**Appendix B: Questions** ...................................    147

**Appendix C: Computations** ................................    149

**Index** ..................................................    155

# Symbols

$\Delta(z)$, 30
$\delta(z, A)$, 24
$\kappa_G$, 27, 88
$\nu$, 3
$\pi(B)$, 32
$\pi[N]$, 33
$v(z, C, \varepsilon)$, 33
$bound(E)$, 69
CF, 2, 3, 5–11, 13–15, 17–20, 22
CS, 4, 5, 8, 17, 18
$\mathrm{dinv}_k(N, d)$, 27
$D_{(m,n)}$, 30
$D_n$, 31
$E(N)$, 24
$E(N, V)$, 24
$E(N, \leq V)$, 24
$E(N, \leq d)$, 24
$E(N, d)$, 24
$frq(b, E)$, 74
FQH, 4
$\mathfrak{G}(N, d, X)$, 27
$G_L$, 88
$\mathrm{GL}(n, k)$, 10
$\mathcal{H}_k(m, N)$, 15
$\mathrm{Inv}_k(N, d)$, 27
IQL, 2–5, 7–13, 15, 18, 19, 21, 22
LL, 2, 6, 14
$\max(A)$, 46
$M(m, n, a, c)$, 33
$M(\mathfrak{n}, a, c)$, 34

$M_0(m,n,a,c)$, 31
MF, 5, 11, 17
MFCF, 2, 3, 6, 7, 11, 17–19, 111
$\mathbb{M}(r,s)$, 61
$\mathbb{M}_+(r,s)$, 61
$\mathbb{M}_2(r,s)$, 61
$p(m,d,n)$, 16
QE, 2, 3, 6, 8–13, 18–20
QH, 2, 3, 6, 9, 13, 18
QP, 2, 4, 6–9, 18
RPA, 2
$symgrp(E)$, 70
$S_N$, 11
$Symgrp(E)$, 70
$Symm_N$, 11
SL$(n,k)$, 12

# Chapter 1
# Fermion Correlations

## 1.1 Introduction

Solid state theory developed from Sommerfeld's realization [1] that simple metals could be described in terms of a gas of free quantum mechanical electrons that obeyed the Pauli exclusion principle [2]. The electrical and thermal conductivities, heat capacity, spin susceptibility and compressibility predicted by this free electron model agreed with experimental observations in simple metals like Al, Na, and K. Early work on the effect of the periodic potential of the solid on the single electron eigenstates [3] led to the concept of energy bands and band gaps, and to some understanding of why some solids were metals, some were insulators, and others were semiconductors [4]. During the early decades of solid state theory, the description of the electronic states rested on the "single particle" picture.

In the middle of the last century, scientists began to question why Sommerfeld's simple model worked so well. The model completely neglected the strong Coulomb interaction $V(r) = \sum_{i<j} e^2 |r_i - r_j|^{-1}$ of the electrons with one another. Treating $V(r)$ in first order perturbation theory added an exchange energy $\mathcal{E}_x(k) = -\left(e^2 k_F/2\pi\right)\left[2 + \frac{k_F^2 - k^2}{k k_F}\ln\left(\frac{k_F+k}{k_F-k}\right)\right]$ to the free electron kinetic energy $\mathcal{E}(k) = \hbar^2 k^2/2m$. In these equations, $k_F$ is the Fermi wave number, and it is given by $k_F = \left(3\pi^2 n_0\right)^{1/3}$, where $n_0$ is the electron density. The exchange energy ruined the agreement with some of the experimental results. In addition, going beyond first order perturbation theory led to divergences. Feynman diagrams depicting the Coulomb interaction to any order in perturbation theory were used in evaluating the ground state energy [5]. Summing certain sets of diagrams to infinite order before integrating over the wave vector removed the divergences that resulted from the long range of the Coulomb interaction. Several authors [6, 7] emphasized the polarization of the electron gas by a moving charged particle, following up on the pioneering work of Lindhard [8, 9]. The "self-energy "of a single excited electron interacting with the polarization cloud that it induces around itself was evaluated [6]. The effec-

© Springer Nature Switzerland AG 2018
S. Mulay et al., *Strong Fermion Interactions in Fractional Quantum Hall States*, Springer Series in Solid-State Sciences 193,
https://doi.org/10.1007/978-3-030-00494-1_1

tive potential $W$ was evaluated in the self–consistent Hartree approximation, and the self-energy given by $\Sigma_0(k, \omega) = G_0(k - k', \omega - \omega') W_0(k', \omega')$. Of course, the right-hand side is to be summed over $k'$ and $\omega'$. $G_0$ is the non-interacting electron propagator (or Greens function) and $W_0$ and $\Sigma_0$ are the effective potential and electron self-energy in lowest non-trivial order. It was demonstrated that the self-consistent Hartree approximation gave the same $W_0$ as the random phase approximation (RPA). Corrections to the RPA, using $G$ in place of $G_0$ and an effective interaction $W$ in place of $W_0$ have been studied [10, 11], but the RPA results are only slightly changed.

Landau [12] had already proposed a phenomenological Fermi liquid theory to describe the effect of short range many-body interactions in liquid $^3$He. The notion of quasiparticles (QPs), elementary excitations that satisfied Fermi–Dirac statistics and included interaction with the ground state, gave rise to important new concepts in solid state theory. Silin [13] made use of Landau's ideas to study the properties of a metallic liquid with long range Coulomb interactions. In all of these approaches, the starting point was still the single particle eigenstates and the Fermi–Dirac distribution function.

During the last two decades, novel systems have been discovered in which many-body interactions appear to dominate over single particle energies. The ultimate example of such a system is the fractional quantum Hall effect. At very large values of a dc magnetic field $B_0$ applied perpendicular to the plane on which the electrons are confined, the massively degenerate single particle Landau levels (LLs) disappear from the problem of determining the ground state and the low energy excitations of the system. Only the Coulomb energy scale $V_c \simeq e^2/\lambda$, where $\lambda = (\hbar c/eB_0)^{1/2}$ is the magnetic length, is relevant to the low energy spectrum. The incompressible quantum liquid (IQL) states discovered by Tsui et al. [14] result from this interaction.

In this chapter, we review the families of IQL states observed experimentally and how they are interpreted. We concentrate on methods that are essential to our explanation, particularly on Laughlin–Jastrow type correlation functions [15] and Jain's mean field composite Fermion (MFCF) picture [16, 17]. The former is obviously a very good approximation for IQL states with filling factor $\nu$ equal to the reciprocal of an odd integer $m$. The validity of Jain's MFCF picture has been justified by our group at the University of Tennessee and its collaborators [18], but only if the interaction energy $V(L_2)$ of a pair of Fermions as a function of the pair angular momentum $L_2 = 2l - R_2$ satisfies certain necessary conditions. Here $l$ is the single particle angular momentum of the Fermion, and $R_2$, the relative pair angular momentum, must be an odd integer. We present a very brief review of Laughlin's remarkable insight into the nature of the correlations giving rise to IQL states, and to his fractionally charged excitations, quasielectrons (QEs) and quasiholes (QHs). We discuss Haldane's idea that the problem of putting fractionally charged quasiparticles (QPs) into a QP Landau level was essentially the same problem as that of putting the electrons into the original electron Landau level. We review Jain's remarkable composite Fermion (CF) picture and demonstrate that it correctly predicts the families of IQL states at filling factors $\nu = n(2pn \pm 1)^{-1}$, where $n$ and $p$ are positive integers. These states correspond to integrally filled composite Fermion Landau levels. The Jain–Laughlin sequence of MFCF states (with $n$ a positive integer) is the most robust set of fractional quantum Hall states observed experimentally. Chen and Quinn [19] introduced an "effective CF angular momentum" $l_0^* = l - p(N - 1)$ associated with the lowest CF Landau level (CFLL0). For $N = 2l_0^* + 1$, this level is exactly filled and a Jain IQL state results. If $N > 2l_0^* + 1$, then $N - (2l_0^* + 1)$ particles must be placed in the next angular momentum shell with $l_1^* = l_0^* + 1$; these are CFQEs. If $N < 2l_0^* + 1$, there will be $2l_0^* + 1 - N$ CFQHs in CFLL0. Thus we have $l_{QH} = l_0^*$ and $l_{QE} = l_1^* = l_0^* + 1$. For any given value of $l$, the single electron angular momentum, one can obtain the number of QEs in the partially filled shell (or the number of QHs in the partially unfilled shell). The lowest band of angular momentum states will contain the minimum number of CFQP excitations consistent with the values of $2l$ and $N$. The value of $(2l, N)$ defines the function space of the $N$ electron system.

## 1.2 The Integral and the Fractional Quantum Hall Effects

The Hamiltonian describing the motion of an electron confined to move on the $xy$-plane in the presence of a dc magnetic field $\boldsymbol{B} = B_0\hat{z}$ perpendicular to the plane is simply $H = (2\mu)^{-1}\left[\boldsymbol{p} + \frac{e}{c}\boldsymbol{A}(\boldsymbol{r})\right]^2$. The vector potential $\boldsymbol{A}(\boldsymbol{r})$ in the symmetric gauge can be taken as $\boldsymbol{A}(\boldsymbol{r}) = \frac{1}{2}B_0\left(-y\hat{x} + x\hat{y}\right)$, giving $\nabla \times \boldsymbol{A} = B_0\hat{z}$. The Schrödinger equation $(H - E)\Psi(\boldsymbol{r}) = 0$ has eigenstates $\Psi_{nm}(r, \phi) = e^{im\phi}u_{nm}(r)$ with eigenvalues $E_{nm} = \frac{1}{2}\hbar\omega_c(2n + 1 + m + |m|)$, where $n \geq 0$ and $m$ are integers and $\omega_c = eB_0/\mu c$ is the cyclotron frequency. The radial wave function is given by $u_{nm}(r) = \chi^{|m|}\exp\left(-\chi^2/2\right)L_n^{|m|}(\chi^2)$, where $\chi^2 = \frac{1}{2}(r/\lambda)^2$, $L_n^{|m|}$ is an associated Laguerre polynomial [18, 20], $Ł_0^{|m|}$ is independent of $\chi$, and $L_1^{|m|}$ is proportional to $(|m| + 1 - \chi^2)$. It is apparent from the eigenvalues that the single particle spectrum consists of highly degenerate levels. The lowest level has $n = 0$ and $m = 0, -1, -2, \ldots$, and its eigenfunction can be written $\Psi_{0m} \propto z^{|m|}\exp\left(-|z|^2/4\lambda^2\right)$, where $z$ stands for $re^{+i\phi}$. For a finite size sample of area $A = \pi\mathcal{R}^2$, the number of single particle states in the lowest Landau level is $N_\phi = B_0A/\phi_0$, where $\phi_0 = \hbar c/e$ is the quantum of flux. The filling factor $\nu$ is defined as $N/N_\phi$, so that $\nu^{-1}$ is simply equal to the number of flux quanta of the dc magnetic field $B_0$ per electron. When $\nu$ is an integer, there is an energy gap between the last filled state and the first empty one. This makes the electron system incompressible, because an infinitesimal decrease in area $A$ can be accomplished only at the expense of promoting an electron across the finite energy gap. This incompressibility is responsible for the integral quantum Hall effect [21, 22]. The energy gaps between the single particle energy levels are the source of the incompressibility.

The observation of an incompressible quantum Hall state in a fractionally filled 2D Landau level [14] was quite unexpected. The behavior of the magneto-resistivity ($\rho_{xx}$ and $\rho_{xy}$) was very similar at filling factor $\nu = 1/3$ to that observed at $\nu = 1, 2, \ldots$. However, there was no gap within the single partial filled Landau level. A gap could result only from the interactions among the electrons. Laughlin [15] proposed that the IQL states observed at filling factor $\nu$ equal to the reciprocal of an odd integer $m$

resulted because all of the electron pairs were able to avoid pair states with relative angular momentum $R_2$ ($R_2 \equiv 2l - L_2$) smaller than $m$ (or separation smaller than $m^{1/2}\lambda$). The avoided pair states had the largest Coulomb repulsion, and avoiding all such states should give an energy minimum. Laughlin proposed a many-body wave function at $\nu = m^{-1}$ given by

$$\Psi_m (1, 2, \ldots, N) = \exp\left(\frac{-\sum_k |z_k|^2}{4\lambda^2}\right) \prod_{i<j} z_{ij}{}^{|m|}. \tag{1.1}$$

For $m = 1$, the product $\prod_{i<j} z_{ij}$ (where $z_i = r_i e^{+i\phi_i}$) is just the Fermion factor which keeps the non-interacting electrons apart. The remaining factor, $G = \prod_{i<j} z_{ij}^{|m-1|}$, is a symmetric correlation factor caused by the Coulomb interactions. It is not difficult to see that $\nu = m^{-1}$, so that $m = 1$ corresponds to a filled LL0, and that $m = 3$ corresponds to a one third filled level. Laughlin also showed that the elementary excitations of the IQL state could be described as fractionally charged QEs and QHs. Both localized and extended states of the quasiparticles were required to understand the behavior of $\rho_{xx}$ and $\rho_{xy}$.

The first explanation of the FQH states at filling factors $\nu = n (1 + 2pn)^{-1}$ with $n > 1$ was given by Haldane [23, 24]. He assumed that the dominant interaction between quasiparticles was the short range part of the pair interaction. If this interaction were sufficiently similar to the Coulomb interaction in electron LL0, the problem of putting $N_{\mathrm{QP}}$ QPs into a QP Landau level would be essentially the same as the original problem of putting $N$ electrons into electron LL0. The number of QP states in the QP Landau level could not exceed $N$, the original number of electrons. This led Haldane to the condition $N = 2pN_{\mathrm{QP}}$ in place of Laughlin's condition $N_\phi = (2p + 1) N$ for the electron IQL states. He picked an even integer $2p$ in place of Laughlin's odd $2p + 1$ because he considers the QPs to be Bosons instead of Fermions. Haldane's hierarchy of IQL states contained all odd denominator fractional filling.

## 1.3   Jain's Composite Fermion Approach

Jain [16, 17] introduced a simple composite Fermion picture by attaching to each electron (via a gauge transformation) a flux tube which carried an even number, $2p$, of magnetic flux quanta. This "Chern–Simons"(CS) flux [25] has no effect on the classical equations of motion since the CS magnetic field $b (r) = 2p\phi_0 \sum_i \delta (r - r_i)$ vanishes at the position $r_i$ of each electron (it is assumed that no electron senses its own Chern–Simons flux). Here $\phi_0$ is the quantum of flux, and the sum is over all electron coordinates $r_i$. This CS transformation results in a much more complicated many-body Hamiltonian which includes a CS vector potential $a (r)$ given by

$$a (r) = \alpha\phi_0 \int d^2 r_i \frac{\hat{z} \times (r - r_i)}{(r - r_i)^2} \psi^\dagger (r_i) \psi (r_i) \tag{1.2}$$

in addition to the vector potential $A(r)$ of the dc magnetic field. In (1.2), $\alpha$ is a constant (it is equal to $2p$ when an even number of flux quanta are attached to each electron), and $\psi^\dagger(r_i)$ (or $\psi(r_i)$) creates (or annihilates) an electron at position $r_i$. The new Hamiltonian (with $A(r)$ replaced by $[A(r) + a(r)]$) simplifies when the mean field (MF) approximation is made. This is accomplished by replacing the density operator $\psi^\dagger(r)\psi(r)$ in the CS vector potential and in the Coulomb interaction by its MF value $n_s$, the uniform MF electron density. The resulting MF Hamiltonian is a sum of single particle Hamiltonians in which an effective magnetic field $B^* = B - \alpha\phi_0 n_s$ appears. The Coulomb interaction disappears because the MF electron charge density $-en_s$ is canceled by the fixed uniform background introduced in order to have charge neutrality. For $\alpha = 2p$, Jain called the particles composite Fermions; they consisted of an electron and the CS flux tube attached to it. In the MF approximation [16, 17], the effective CF filling factor $\nu^*$ satisfied the equation $(\nu^*)^{-1} = \nu^{-1} - 2p$, i.e., the number of flux quanta per electron due to the dc magnetic field $B_0$ less the CS flux per electron introduced in the CS transformation. When $\nu^*$ is equal to an integer $n = \pm 1, \pm 2, \ldots$, then $\nu = n(1 + \alpha n)^{-1}$. For $\alpha = 2$, this generates IQL Hall states at $\nu = {}^1/_3, {}^2/_5, {}^3/_7, \ldots$ and $\nu = 1, {}^2/_3, {}^3/_5, \ldots$. These are the most prominent FQH states observed experimentally, and they correspond to integrally filled CF Landau levels.

It is convenient to take the 2D surface on which the electrons reside to be a sphere of radius $\mathcal{R} = (A/2\pi)^{1/2}$ with a magnetic monopole of strength $2Q$ flux quanta at its center causing a radial magnetic field of magnitude $B_0 = 2Q(\hbar c/e)/(4\pi R^2)$. This spherical geometry [23, 24] has the advantage of a finite surface area with full rotational symmetry. The single particle angular momentum $l$ has a projection $l_z$ satisfying $-l \le l_z \le l$. On the plane $z = 0$, the allowed values of $m_z$, the $z$-component of angular momentum, must belong to the set $g_0 = \{0, 1, 2, \ldots, 2l\}$. The total angular momentum $L$ of the $N$ particle system has a projection $L_z$ on the sphere. On the plane, $M$ is defined as the sum over all electrons of the value of $m_z$ for each electron. Because $l_z$ and $m_z$ differ by $l$, it is clear $L_z = M - Nl$. The eigenstates on the sphere can be written as $|L, L_z\rangle$ and on the plane as $|M_R, M_{CM}\rangle$, where $M = M_R + M_{CM}$ is the sum of relative and center of mass angular momenta. Interaction energies depend only on $L$ but not $L_z$, and only on $M_R$ but not $M_{CM}$ [18]. It is apparent that $M_R = Nl - L$ and $M_{CM} = L + L_z$. To construct an $N$ electron product state of angular momentum $L = 0$, a linear combination of product states with $L_z = 0$ is required, implying that $L = Nl - M$.

As noted earlier [19], Chen and Quinn introduced an effective CF angular momentum $l_0^*$ satisfying the relation $l_0^* = l - p(N-1)$, where $2p$ is the number of flux quanta attached to each electron in the CS transformation. In the spherical geometry, this results from taking an effective monopole strength, $2Q^* = 2Q - 2p(N-1)$, seen by each composite Fermion. Then $Q^* = l_0^*$ is the CF angular momentum. For CFLL0 (lowest CF Landau level) filled and CFLL1 completely empty, $2l_0^* = N - 1$, and an integrally filled CF state results. If $2l_0^* \ne N - 1$, CFQEs of angular momentum $l_{QE} = l_0^* + 1$ (or CFQHs of angular momentum $l_{QH} = l_0^*$) occur. In Table 1.1, we give the values [18] of $2Q^*$, $n_{QH}$, $n_{QE}$, $l_{QH}$, $l_{QE}$, and $L$ for a system of $N = 10$

**Table 1.1** The effective CF Monopole strength $2Q^*$, the number of CF quasiparticles (quasiholes $-n_{QH}$ and quasielectrons $n_{QE}$), the quasiparticle angular momenta $l_{QH}, l_{QE}$ and the angular momenta $L$ of the lowest lying band of multiplets for a ten electron system at $2Q$ from 25 to 29

| $2Q$ | 29 | 28 | 27 | 26 | 25 |
|---|---|---|---|---|---|
| $2Q^*$ | 11 | 10 | 9 | 8 | 7 |
| $n_{QH}$ | 2 | 1 | 0 | 0 | 0 |
| $n_{QE}$ | 0 | 0 | 0 | 1 | 2 |
| $l_{QH}$ | 5.5 | 5 | 4.5 | 4 | 3.5 |
| $l_{QE}$ | 6.5 | 6 | 5.5 | 5 | 4.5 |
| $L$ | 10, 8, 6, 4, 2, 0 | 5 | 0 | 5 | 8, 6, 4, 2, 0 |

electrons when $2Q$ ranges from 25 to 29. The total angular momentum $L$ is obtained by addition of $QP$ angular momenta of $n_{QP}$ quasiparticles treated as Fermions.

In the mean field approximation, the CFQPs do not interact with one another. Therefore, the states with two QEs ($L = 0 \oplus 2 \oplus 4 \oplus 6 \oplus 8$) and with two QHs ($L = 0 \oplus 2 \oplus 4 \oplus 6 \oplus 8 \oplus 10$) should form degenerate bands. Numerical diagonalization of the ten electron system clearly shows that the two quasiparticles states are not degenerate. The deviation of the energies $E(L)$ from their non-interacting energy $E_0 = 2\mathcal{E}_{QP}$ gives their pair interaction energy as a function of their pair angular momentum $L_2$ up to a constant which does not influence correlations. Figure 1.1 shows $E(L)$, the energy as a function of total angular momentum for the ten electron system. It is apparent that Jain's MFCF picture gives the correct values of $L$ for the lowest band of states obtained by exact numerical diagonalization (within the subspace of the partially filled LL).

For large systems (e.g., $N > 14$), numerical diagonalization of the electron-electron interactions becomes difficult, so we have investigated the low lying energy states by determining the number of QEs and QHs ($n_{QE}$ or $n_{QH}$), their angular momenta $l_{QE}$ and $l_{QH}$, and their interaction energies $V_{QE}(L_2)$ and $V_{QH}(L_2)$. Since $n_{QE}$ (or $n_{QH}$) is much smaller than $N$, and $l_{QE}$ (and $l_{QH}$) much smaller than $l$, the electron angular momentum, we can easily diagonalize these smaller systems. One example [18] is shown in Fig. 1.2 for the case $(2l, N) = (29, 12)$, which corresponds to $(2l_{QE}, n_{QE}) = (9, 4)$. The low lying states of the electron system are very close to those of the four QE system, suggesting that description in terms of QP excitations interacting via $V_{QP}(L_2)$ is reasonable.

## 1.4   The Composite Fermion Hierarchy

Sitko et al. [26] introduced a very simple CF hierarchy picture in an attempt to understand Haldane's hierarchy of Laughlin correlated daughter states and Jain's sequence of IQL states with integrally filled CF LLs. Jain's MFCF picture neglected interac-

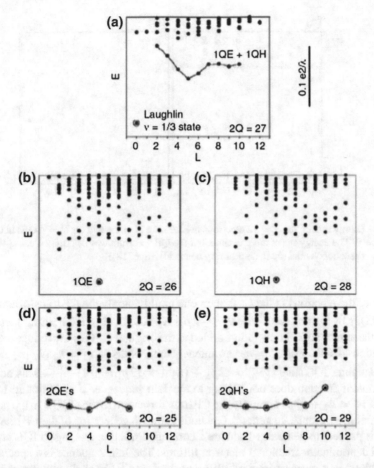

**Fig. 1.1** The spectra of ten electrons in the lowest Landau level calculated on a Haldane sphere with $2Q$ from 25 to 29. The open circles and solid lines mark the lowest energy bands with the fewest composite fermion quasiparticles. (See [18], for example, and references therein)

tions between QPs. The gaps causing incompressibility were energy separations between the filled and lowest empty single particle CF LLs. Not all odd denominator fractions occurred in the Jain sequence $\nu = n(2pn \pm 1)^{-1}$, where $n$ and $p$ are non-negative integers. The missing IQL states were ones with partially filled CF QP shells. The energy gap causing their incompressibility resulted from *residual interactions* between the CF QPs. For an initial electron filling factor $\nu_0$, the relation between $\nu_0$ and $\nu_0^*$, the effective CF filling factor, satisfied $\nu_0^{-1} = (\nu_0^*)^{-1} + 2p_0$, and gave rise to the Jain states when $\nu_0^*$ was equal to an integer $n$. What happens if $\nu_0^*$ is not an integer? It was suggested [26] that then one could write $\nu_0^* = n_1 + \nu_1$, where $n_1$ was an integer and $\nu_1$ represented the filling factor of the partially filled CF QP shell. If Haldane's assumption that the pair interaction energy $V_{QP}(L_2)$, as a function of the angular momentum $L_2$ of the QP pair, was sufficiently similar to $V_0(L_2)$, the interaction

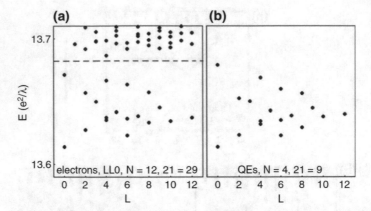

**Fig. 1.2** Energy spectra for $N = 12$ electrons in LL0 with $2l = 29$, and for $N = 4$ QEs in CF LL1 with $2l = 9$. The energy scales are the same, but the QE spectrum was determined using $V_{QE}(\mathcal{R})$ as the pair psuedopotential (up to an arbitrary constant). (See [18])

energy of the electrons in the LL0, then one could reapply the CF transformation to the CF QPs by writing $\left(\nu_1^*\right)^{-1} = \nu_1^{-1} - 2p_1$. Here, $\nu_1$ is the CF QP filling factor and $2p_1$ is the number of CS flux quanta added to the original CF QPs to produce a second generation of CFs. For $\nu_1^* = n_2$, an integer, this results in $\nu_1 = n_2 (2p_1n_2 \pm 1)^{-1}$, and a daughter IQL state at $\nu_0^{-1} = 2p_1 + \left[n_1 + n_2(2p_1n_2 + 1)^{-1}\right]^{-1}$. This new odd denominator fraction does not belong to the Jain sequence. If $\nu_1^*$ is not an integer, then set $\nu_1^* = n_2 + \nu_2$ and reapply the CF transformation to the CF QE in the new QP shell of filling factor $\nu_2$. In general, one finds at the $l$th generation of the CF hierarchy, that this procedure generates Haldane's continued fraction leading to IQL states at all odd denominator fractional electron fillings. The Jain sequence is a special case in which $\nu_0^* = n$ gives an integral filling of the first CF QP shell, and the gap is the separation between the last filled and first empty CF levels.

The CF hierarchy picture was tested by Sitko et al. for the simple case of $(2l, N) = (18, 8)$ for LL0 by comparing its prediction to the result obtained through exact numerical diagonalization. For this case, $2l_0^* = 2l - 2(N - 1) = 18 - 2(7) = 4$. Therefore, CF LL0 can accommodate $2l_0^* + 1 = 5$ CFs. The three remaining CFs must go into CF LL1 as CF QEs of angular momentum $l_{QE} = l_0^* + 1 = 3$. This generates a band of states with $L = 0 \oplus 2 \oplus 3 \oplus 4 \oplus 6$. This is exactly what is found for the lowest energy band of states obtained by numerical diagonalization shown in Fig. 1.3. Reapplying the CF transformation to the first generation of CF QEs would generate $2l_1^* = 2l_0^* - 2(n_{QE} - 1) = 4 - 2(2) = 0$, giving an $L = 0$ daughter IQL state if the CF hierarchy were correct. Clearly, the lowest energy state obtained in the numerical diagonalization does not have angular momentum $L = 0$ as predicted by the CF hierarchy. The $L = 0$ and $L = 3$ multiplets clearly have higher energies than the other three multiplets. Sitko et al. conjectured that this must have resulted because the psuedopotential $V_{QE}(L_2)$ was not sufficiently similar to that of electrons in LL0

**Fig. 1.3** Low energy spectrum of 8 electrons at $2l = 18$. The lowest band contains 3 QEs, each with $l_{QE} = 3$. Reapplying the CS mean-field approximation to these QEs would predict an $L = 0$ daughter state corresponding to $\nu = {}^4/_{11}$. The data makes it clear that this is not valid. (See [26])

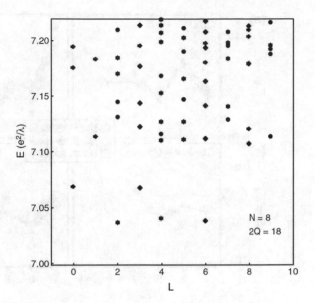

to support Laughlin correlations. Laughlin correlations are essential for forming a next generation of CFs.

The QEs and QHs have residual interactions that are more complicated than the simple Coulomb interaction in LL0. We have already seen, from Fig. 1.1d and e, that we can obtain $V_{QP}(L_2)$ up to an overall constant from numerical diagonalization of $N$-electron systems in LL0. More careful estimates of $V_{QE}(\mathcal{R})$ and $V_{QH}(\mathcal{R})$ (where $\mathcal{R} = 2l - L_2$, and $L_2$ is the pair angular momentum) are shown in Fig. 1.4. We define a psuedopotential to be *harmonic* if it increases with $L_2$ as $V_H(L_2) = A + BL_2(L_2 + 1)$, where $A$ and $B$ are constants. For LL0, the actual psuedopotential $V(L_2)$ always increases with $L_2$ more rapidly than $V_H(L_2)$. For QEs in CF LL1, the psuedopotential $V_{QE}(L_2)$ has minima at $L_2 = 2l - 1$ and at $L_2 = 2l - 5$, and a maximum at $L_2 = 2l - 3$. This oscillatory behavior of the interaction energy of a QE pair must be responsible for the failure of the CF hierarchy prediction of an $L = 0$ IQL state.

## 1.5 Justification of the CF Approach

Pan et al. [27] found IQL states of electrons in LL0 that do not belong to the Jain sequence of integrally filled CF states. One example is the $\nu = {}^4/_{11}$ filling factor of a state that is assumed to be fully spin polarized. Numerical diagonalization studies of fully spin polarized systems did not find an $L = 0$ IQL ground state at $\nu_{QE} = {}^1/_3$, which would result in an IQL state at electron filling factor of $\nu = {}^4/_{11}$. In addition, Pan et al. found strong minima in $\rho_{xx}$ at even denominator filling factors

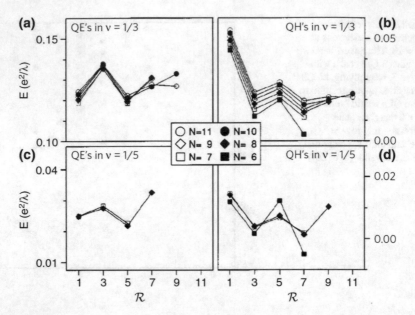

**Fig. 1.4** The pseudopotentials of a pair of quasielectrons (left) and quasiholes (right) in Laughlin $\nu = {}^1/_3$ (top) and $\nu = {}^1/_5$ (bottom) states, as a function of relative angular momentum $\mathcal{R}$. Different symbols mark data obtained in the diagonalization of from 6 to 11 electrons

($\nu = {}^3/_8$ and $\nu = {}^3/_{10}$) suggesting the existence of IQL states that can't be part of the CF hierarchy. Our research group has made an important contribution to this field, by rigorously proving [28–32] under which conditions Jain's elegant CF approach correctly predicts the angular momentum multiplets belonging to the lowest energy sector of the spectrum for any value of the applied magnetic field. Because there is no small parameter in this strongly interacting many-body system, our proof does not involve treating fluctuations beyond the MF by a perturbation expansion. It involves proving some rigorous mathematical theorems and applying them, together with well-known concepts frequently used in atomic and nuclear physics. We do not review the arguments on why and when Jain's MFCF picture correctly predicts the angular momentum multiplets in the lowest band but urge the interested reader to look at [28–32]. Our rigorous theorem that should be emphasized states that for a harmonic pair psuedopotential $V_H(L_2) = A + BL_2(L_2 + 1)$, every multiplet $\left| l^N, L\alpha \right\rangle$ with the same value of total angular momentum $L$ has the same energy,

$$
E_\alpha(L) = N\left[\frac{1}{2}(N-1)A + B(N-2)l(l+1)\right] \\
+ BL(L+1),
\tag{1.3}
$$

independent of multiplet index $\alpha$. This means that a harmonic psuedopotential $V_H(L_2)$ does not cause correlations (i.e., does not remove the degeneracy of mul-

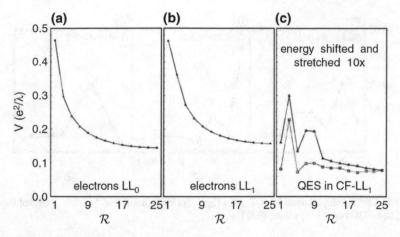

**Fig. 1.5** Pair interaction pseudopotentials as a function of relative angular momentum $\mathcal{R}_2$ for electrons in the LL0 (**a**), LL1 (**b**) and for the QEs of the Laughlin $\nu = {}^1/_3$ state by Lee et al. [33, 34] (squares), and by Wójs et al. [29, 30] (triangles). (See [18])

tiplets having the same value of total angular momentum $L$). Only the deviation $\Delta V (L_2)$ defined by $\Delta V (L_2) = V (L_2) - V_H (L_2)$ results in correlations. The simplest model for $\Delta V (L_2)$ is one with only short range anharmonic behavior,

$$\Delta V (L_2) = k\delta (L_2, 2l - 1). \tag{1.4}$$

If $k > 0$, it is apparent that the lowest energy multiplet for each value of total angular momentum $L$ is the one which avoids (to the maximum possible extent) having pairs with $L_2 = L_2^{MAX} = 2l - 1$. This is just what is meant by Laughlin correlations, and is the reason why the Laughlin trial wave function for the $\nu = {}^1/_3$ filled state is the exact solution to a short range pair interaction psuedopotential. If $k < 0$, then the lowest energy state (when $\nu$ is not too small) for each value of $L$ is the one with $P_{L\alpha} (L_2^{MAX})$, the probability that $\left| l^N, L\alpha \right\rangle$ has pairs with pair angular momentum $L_2 = L_2^{MAX}$, having a maximum value. This corresponds to forming pairs with $L_2 = L_2^{MAX} = 2l - 1$. It is important to emphasize that $V (L_2)$ rises faster than $V_H (L_2)$ at all values of $L_2$ for electrons in LL0. In LL1, this is not true for $2l - 1 \geq L_2 \geq 2l - 5$. Thus in LL1, we do not expect Laughlin correlations for filling factor $\nu = {}^1/_3$.

In Fig. 1.5, we display $V (\mathcal{R}_2)$, where $\mathcal{R}_2 = 2l - L_2$ for a) electrons in LL0, b) electrons in LL1, and c) QEs in CFLL1. It is clear that for QEs of the $\nu = {}^1/_3$ filled IQL state, $V_{QE} (L_2)$ is not increasing with $L_2$ faster than the harmonic psuedopotential. In fact, unlike $V_0 (L_2)$ and $V_1 (L_2)$, it is not even a monotonically increasing function of $L_2$. Therefore, we certainly do not expect Laughlin correlations among the CFQEs in CFLL1.

Moore and Read [35] treated the $\nu = {}^5/_2$ state (thought of as $\nu = 2 + {}^1/_2$) as a half filled spin polarized state of LL1 with both spin states of LL0 occupied. They suggested that the state had "pairing correlations" similar to those occurring in

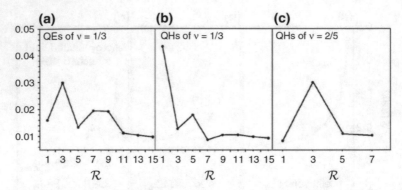

**Fig. 1.6** The pseudopotentials $V_{QE}(\mathcal{R}_2)$ and $V_{QH}(\mathcal{R}_2)$ for **a** QEs of $\nu = {}^1/_3$ state, **b** QHs of $\nu = {}^1/_3$ state, and **c** QHs of $\nu = {}^2/_5$ state. (See [19])

superconductors. We expect the QEs of the Laughlin $\nu = {}^1/_3$ state to form pairs because $V_{QE}(L_2)$ has a maximum at $\mathcal{R}_2 \equiv 2l - L_2$ equal to 3 and 7, and a minimum at $\mathcal{R}_2 = 1$ and 5. A schematic of $V_{QE}(\mathcal{R}_2)$ vs $\mathcal{R}_2$ is shown in Fig. 1.6a for QEs of the $\nu = {}^1/_3$ Laughlin IQL, (b) for QHs of the $\nu = {}^1/_3$ Laughlin state, and (c) for QHs of the $\nu = {}^2/_5$ Jain state.

The simplest way of picturing paired states is to introduce a pair angular momentum $l_P = 2l - 1$ and form $N_P = N/2$ such pairs. The pairs cannot get too close to one another without violating the Pauli principle. One would normally think of pairs of Fermions as Bosons, but in two dimensional systems we can alter the particle statistics by using a Chern–Simons transformation. We introduce a Fermion pair (FP) angular momentum $l_{FP}$ satisfying the equation

$$2l_{FP} = 2l_P - \gamma_F (N_P - 1).$$ (1.5)

For a single pair, $l_{FP} = 2l - 1$. As $N_P$ increases, the allowed values of the total angular momentum are restricted to values less than or equal to $2l_{FP}$. The value of the constant $\gamma_F$ is determined by requiring that the FP filling factor be equal to unity when the single Fermion filling factor has an appropriate value. For $l_P = 2l - 1$, this value corresponds to single Fermion filling $\nu = 1$. Setting $\nu_{FP}^{-1} = (2l_{FP} + 1)/N_P$, $\nu^{-1} = (2l + 1)/N$, and $N_P = N/2$ gives $\nu_{FP}^{-1} = 4\nu^{-1} - 3$, (i.e., $\gamma_F = 3$), so that $\nu_{FP} = 1$ when $\nu = 1$. The factor of 4 multiplying $\nu^{-1}$ results from the pairs having a charge of $-2e$, and $N_P$ being equal to $N/2$. This procedure allows the Fermion pairs to be Laughlin correlated instead of the individual electrons being so. It predicts that an IQL state at $\nu_1 = {}^1/_2$ occurs when $2l = 2N - 3$ as found by Moore and Read.

We have applied the same idea to the quasielectrons and quasiholes of the Laughlin $\nu = {}^1/_3$ state, and to quasiholes of the Laughlin–Jain $\nu = {}^2/_5$ state. QHs of $\nu = {}^1/_3$ state reside in CF LL0, but both QEs of the $\nu = {}^1/_3$ state and the QHs of the $\nu = {}^2/_5$ state reside in CF LL1.

If we assume that the QEs form pairs and treat the pairs as Fermions, then (1.5) gives the relation between the *effective FP angular momentum* $l_{FP}$, and the QE angular momentum $l$, and the relation between the *effective FP filling factor* $\nu_{FP}$, and the QE filling factor $\nu_{QE}$. If we take $\nu_{FP} = m^{-1}$, when $m$ is an odd integer, we can obtain the value of $\nu_{QE}$ corresponding to the Laughlin correlated state of FPs (pairs of quasielectrons with $l_P = 2l - 1$). Exactly the same procedure can be applied to QHs in CF LL1 since $V_{QE}(\mathcal{R})$ and $V_{QH}(\mathcal{R})$ are qualitatively similar at small values of $\mathcal{R}$. Here, we are assuming that $V_{QE}(\mathcal{R})$ and $V_{QH}(\mathcal{R})$ are dominated by their short range behavior $\mathcal{R} \leq 5$. The QH psuedopotential is not as well determined for $\mathcal{R} > 5$ because it requires larger $N$ electron systems than we can treat numerically. The electron filling factor is given by $\nu^{-1} = 2 + (1 + \nu_{QE})^{-1}$ or by $\nu^{-1} = 2 + (2 - \nu_{QH})^{-1}$. This results in the values of $\nu$ in qualitative agreement with experimental results.

## 1.6 Numerical Diagonalization Studies

Confirmation of the Laughlin explanation of when IQL ground states occur can be found through numerical diagonalization of the Coulomb interaction between electrons. For $\hbar\omega_c$ large compared to the Coulomb energy scale $V_c \simeq e^2/\lambda$, only the subspace of the lowest Landau level (LL0) is relevant to determining the low energy states in the spectrum. Numerical diagonalization is usually performed in Haldane's spherical geometry. A concise explanation of this set-up is as follows. There is a one-to-one correspondence between $N$ electrons in a plane described by coordinates $(r, \phi)$ and the $N$ electrons on a sphere described by $(l, l_z)$. For the plane, the $z$-component of angular momentum takes on the values $m = 0, 1, \ldots, N_\phi$ and the total $z$-component of angular momentum is $M = \sum_{i=1}^{N} m_i$, where $m_i$ is the $z$-component of the angular momentum of particle $i$. $M$ is the sum of the relative angular momentum $M_R$ and the center of mass angular momentum $M_{CM}$. On a sphere, the $z$-component of the single particle angular momentum is written as $l_z$, and $|l_z| \leq l$, where $l$ is the angular momentum in the shell (or Landau level). The total angular momentum $L$ is the sum of the angular momenta of $N$ Fermions, each with angular momentum $l$. $N$ electron states are designated by $|L, L_z, \alpha\,\alpha\rangle$, where $\alpha$ is used to label different multiplets with the same value $L$. It is apparent that $M = Nl + L_z$ and one can show that $M_R = Nl - L$ and $M_{CM} = L + L_z$. Therefore, for a state of angular momentum $L = 0$, $M_R$ must be equal to $Nl$. In general, the value of $L$ for a given correlation function $\mathcal{G}$ is determined by the equation $L = Nl - K_\mathcal{F} - K_\mathcal{G}$, where $K_\mathcal{F} = N(N - 1)/2$ is the number of CF-lines appearing in the Fermi function $\mathcal{F}$ and $K_\mathcal{G}$ is the number of CF-lines appearing in the correlation function $\mathcal{G}$. For $L = 0$, this relation was given by [36]. In Haldane's spherical geometry, the $N$ electron system is confined to move on the surface of a sphere of radius $R$, with a magnetic monopole of strength $2Q$ flux quanta sitting at the center ($2Q$ is taken to be an integer). This results in a radial magnetic field of magnitude $B_0 = 2Q\phi_0/4\pi R^2$. This geometry avoids boundary conditions, has full rotational symmetry, and contains a finite number of particles. The single particle eigenfunctions are called monopole harmonics [37] and can be written as $\left|Qlm\right\rangle$. They are eigenfunctions of the single particle Hamiltonian $\hat{H}_0$, the square of the single particle angular momentum $\hat{l}$, and its $z$-component $\hat{l}_z$. The eigenvalues are given by $E_{Qlm} = \left(\hbar\omega_c/2Q\right)\left[l(l+1) - Q^2\right]$, $l(l+1)$, and $m$, respectively. The lowest Landau level (or angular momentum shell) has $l$ equal to $Q$, half the monopole strength, so LL0 has angular momentum $l_0 = Q$ and the excited LLs have $l_n = Q + n$. The angular momentum shell contains $2l+1$ states with $-l \leq m \leq l$. Note that $N$ particle Fermion states can be written $\left|m_1, m_2, \ldots, m_N\right\rangle = C_N^\dagger C_{N-1}^\dagger \ldots C_1^\dagger\left|0\right\rangle$, where $m_i$ belongs to the set $g_0$ with $|m| \leq l$. Here, $C_j^\dagger$ creates an electron with $z$ component of angular momentum equal to $m_j$, and $\left|0\right\rangle$ is the ket for the vacuum state. The interaction Hamiltonian is $H_I = \sum_{i,j} e^2/r_{ij}$, and matrix elements $\left\langle m_1' m_2' \cdots m_N' \right| H_I \left| m_1 m_2 \cdots m_N \right\rangle$ vanish unless

(i) $M = \sum_i m_i' = \sum_i m_i$, where the sum is over all occupied states, and
(ii) $\left| m_1 m_2 \cdots m_N \right\rangle$ and $\left| m_1' m_2' \cdots m_N' \right\rangle$ differ by no more than two members.

The spherical symmetry allows use of the Wigner–Eckart theorem

$$\left\langle L'M'\alpha' \right| H_I \left| LM\alpha \right\rangle = \delta_{LL'}\delta_{MM'}\left\langle L\alpha' \right| H_I \left| L\alpha \right\rangle, \tag{1.6}$$

and the reduced matrix element on the right hand side is independent of $M$. The matrices to be diagonalized are very large but sparse, and standard programs allow diagonalization for $N$ up to roughly 12 or 14. The numerical results displayed in Figs. 1.1 and 1.2 are representative of the general results obtained for $N \leq 14$. It should be noted in Fig. 1.1 that there is a gap between the lowest state or band of states and a quasi continuum above them. It should also be stressed that the spread in values of energy for the 2QP bands (frames d and e) is small compared to this gap.

A reasonable understanding of the ground states and elementary excitations of the states in the Laughlin–Jain sequence for partially filled LL0 results from the agreement of the Laughlin–Jain correlations with the numerical results obtained by diagonalization within the partially occupied Landau level.


## 1.7   Correlations and Correlation Diagrams

Laughlin [15] realized that if the interacting electrons could avoid the most strongly repulsive pair states, an incompressible quantum liquid (IQL) state could result. He suggested a trial wave function for a filling factor $\nu$ equal to the reciprocal of an odd integer $n$, in which the correlation function $G_n\left(z_{ij}\right)$ was given by $\prod_{i<j} z_{ij}^{n-1}$. This function is symmetric and avoids all pair states with relative pair angular momentum smaller than $n$ (or all pair separations smaller than $r_n = n^{1/2}\lambda$). One can represent this Laughlin correlation function $G_L$ diagrammatically by distributing $N$ dots, representing $N$ electrons on the circumference of a circle, and drawing double lines, representing two correlation factors (cfs) connecting each pair. There are $2(N-1)$

CF factors in $G_L \{z_{ij}\}$ emanating from each particle $i$. Adding $(N - 1)$ CF factors emanating from each particle due to the Fermion factor $F \{z_{ij}\}$ gives a total of $3(N - 1)$ cfs emanating from each particle in the trial wave function $\Psi$. This number cannot exceed $N_\phi = 2l$ defining the function space $(2l, N)$ of the LL0.

The other well-known trial wave function is the Moore–Read [35] paired function describing the IQL state of a half filled spin polarized first excited Landau level (LL1). This wave function $\Psi$ can be written in the form $\Psi = F \cdot G_{MR}$, where the correlation function is taken as $G_{MR} = F \{z_{ij}\} Pf \left( z_{ij}^{-1} \right)$. The second factor is called the Pfaffian of $z_{ij}^{-1}$. It can be expressed as [25, 35]

$$Pf \left( z_{ij}^{-1} \right) = \hat{A} \left( \prod_{i=1}^{N/2} (z_{2i-1} - z_{2i})^{-1} \right), \tag{1.7}$$

where $\hat{A}$ is an antisymmetrizing operator and the product is over pairs of electrons. There has been considerable interest in the Moore–Read paired state and its generalizations [38–40] based on rather formidable conformal field theory. In [41], we propose a simple intuitive picture of Moore–Read correlations with the hope that it might lead to new insight into correlations in strongly interacting many-body systems (see Appendix A for a detailed treatment).

For the simple case of an $N = 4$ particle system, the Pfaffian can be expressed as

$$Pf \left( z_{ij}^{-1} \right) = \hat{A} \{ (z_{12}z_{34})^{-1} \}$$
$$= \left[ (z_{12}z_{34})^{-1} - (z_{13}z_{24})^{-1} + (z_{14}z_{23})^{-1} \right]. \tag{1.8}$$

The product of $F \{z_{ij}\}$ and $Pf \left( z_{ij}^{-1} \right)$ gives for the Moore–Read correlation function

$$G_{MR} \{z_{ij}\} = z_{13}z_{14}z_{23}z_{24} - z_{12}z_{14}z_{23}z_{34}$$
$$+ z_{12}z_{13}z_{24}z_{34}. \tag{1.9}$$

The correlation diagram for $G_{MR} \{z_{ij}\}$ contains four points with a pair of cfs emanating from each particle $i$ going to different particles $j$ and $k$. There are three distinct diagrams shown in Fig. 1.7. Note that $G_{MR}$ is symmetric under permutation, as it must be, since it is a product of two antisymmetric functions $F \{z_{ij}\}$ and $Pf \left( z_{ij}^{-1} \right)$.

A simpler, but seemingly different, correlation is the quadratic function given by $G_Q \equiv \hat{S} \left( z_{12}^2 z_{23}^2 \right)$, where $\hat{S}$ is a symmetrizing operator. The correlation diagram for $G_Q \{z_{ij}\}$ is shown in Fig. 1.8. $G_{MR}$ and $G_Q$ are clearly different. However, when they are expressed as homogeneous polynomials in the independent variables $z_1$ to $z_4$ (by simple multiplication), the two polynomials are the same up to normalization constant. The same was true for an $N = 6$ particle system, leading to the conjecture that $G_{MR} \{z_{ij}\}$ was equivalent to $G_Q \{z_{ij}\}$ for all $N$ [38]. This conjecture was proved

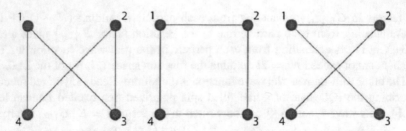

**Fig. 1.7** Moore–Read correlation diagram for $N = 4$. Dots represent the particles, and solid lines the cfs $z_{ij}$. $G_{MR}$ is the symmetric sum of the three diagrams and is given by (1.9)

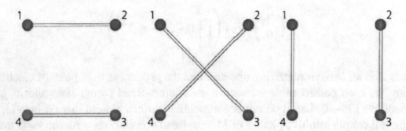

**Fig. 1.8** Quadratic correlation functions. The square of a CF, $z_{ij}^2$, is represented by double lines. $G_Q$ is the sum of the contributions from the three diagrams

by our group before we discovered that Cappelli et al. [39] had already shown the equivalence.

There are several advantages to the use of $G_Q$. First, it is simpler to partition $N$ into two subsets $A$ and $B$ of size $N/2$, (e.g., $A = \{1, 2, \ldots, N/2\}$ and $B = \{N/2 + 1, \ldots, N\}$), and define $g_{AB} = g_A g_B = \prod_{i<j\in A} z_{ij}^2 \prod_{k<l\in B} z_{kl}^2$ for each $(A, B)$. Then the full correlation function can be written as $\hat{S}_N \{g_{AB}\}$, where $\hat{S}_N$ symmetrizes $g_{AB}$ over all $N$ particles. This symmetrization is equivalent to summing $g_{AB}$ over all possible partitions of $N$ into two equal size subsets $A$ and $B$. In Fig. 1.9, we show the contribution to $G_Q$ for $N = 8$ particles for a partition in which $A = \{1, 3, 5, 7\}$ and $B = \{2, 4, 6, 8\}$.

Jain [16, 17] introduced a composite Fermion (CF) picture by attaching to each electron (via a gauge transformation) a flux tube which carried an even number $2p$ of magnetic flux quanta. This *Chern–Simons* (CS) flux has no effect on the classical equations of motion since the CS magnetic field $b(r) = 2p\phi_0 \sum_i \delta(r - r_i) \hat{z}$ vanishes at the position of each electron (it is assumed that no electron senses its own CS flux). Here, $\phi_0$ is the quantum of flux, and the sum is over all electron coordinates $r_i$. The classical Lorentz force on the $i$th electron due to the CS magnetic field is $(-e/c) v_i \times b(r_i)$, and $b(r_i)$ caused by the CS flux on every $j$ not equal to $i$ vanishes at the position $r_i$. The CF model results in a much more complicated interaction Hamiltonian, but simplification results from making a mean field (MF) approximation in which the CS flux and the electron charge are uniformly distributed over the

**Fig. 1.9** Correlation
diagram for $G_Q\{z_{ij}\}$ in an
eight electron system due to
the partition $A = \{1, 3, 5, 7\}$
and $B = \{2, 4, 6, 8\}$. The full
correlation function is the
sum over all distinct
partitions of $N$ into subsets
$A$ and $B$ each containing
$N/2 = 4$ particles. The trial
wave function is
$\Psi_Q(1, 2, \ldots, 8) =$
$F\{z_{ij}\} G_Q\{z_{ij}\}$

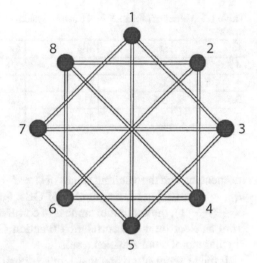

entire sample. The average electronic charge $-eN/\mathcal{A}$ is canceled by the fixed background of positive charge introduced to make the total charge vanish. This MFCF approximation results in a system of $N$ non-interacting CFs (CF = electron plus attached flux tube) moving in an effective magnetic field $b^* = \nu b$. An effective CF filling factor $\nu^*$ was introduced satisfying the equation

$$\left(\nu^*\right)^{-1} = \nu^{-1} - 2p. \tag{1.10}$$

This resulted in a filled CF level when $\nu^*$ was equal to an integer ($\nu^* = n = \pm 1, \pm 2, \ldots$) and a IQL daughter state at $\nu = n(1 + 2pn)^{-1}$. This Jain sequence of states was the most robust set of fractional quantum Hall states observed in experiments.

Making use of Haldane's spherical geometry [23, 24, 37, 38], Chen and Quinn [19] introduced an *effective CF angular momentum* $l^*$ satisfying the relation $l_0^* = l - p(N - 1)$, where $2p$ is the number of CS flux quanta per electron. The lowest CF Landau level (CF LL0) could hold $(2l^* + 1)$ CFs. There would be $n_{QE} = N - (2l^* + 1)$ composite Fermion QEs of angular momentum $l_{QE} = l^* + 1$, or $n_{QH} = (2l^* + 1) - N$ CF QHs of angular momentum $l_{QH} = l^*$ if $2l^* + 1$ was not equal to $N$. This resulted in a lowest band of quasiparticle (QP) states separated by a gap from the higher energy quasi continuum. This allowed the total angular momentum states in this band to be determined by the addition of angular momenta of $n_{QP}$ quasiparticles each of angular momentum $l_{QP}$ according to addition of Fermion angular momenta.

In Table 1.2, we summarize the results of Jain's MFCF picture of the low energy states of an $N = 4$ electron system for values of $2l$ equal to 9, 8, 7, and 6. These correspond to the $\nu = {}^1/_3$ filled IQL states and its excited states containing one, two, and three QEs. The table gives the values of $l$, the single electron angular

**Table 1.2** Values of $l$ for an $N = 4$ electron system and the values of $l_0^*, n_{QE}, l_{QE}, k_M$, and $L$ which result

| $l$ | $l_0^*$ | $n_{QE}$ | $l_{QE}$ | $k_M$ | $L$ |
|-----|---------|----------|----------|-------|-----|
| 4.5 | 1.5 | 0 | 2.5 | 6 | 0 |
| 4 | 1 | 1 | 2 | 5 | 2 |
| 3.5 | 0.5 | 2 | 1.5 | 4 | $0 \oplus 2$ |
| 3 | 0 | 3 | 1 | 3 | 0 |

momentum, and the resulting values of $l_0^* = l - (N - 1)$, the CF angular momentum; $n_{QE} = N - (2l_0^* + 1)$, the number of QEs; $l_{QE}$, the QE angular momentum; $k_M = 2l - (N - 1)$, the maximum number of correlation factor (CF) lines that can emanate from an electron in the correlation function $G$, and the allowed values of the total angular momentum $L$ which result.

It might seem surprising that Jain's very simple CF picture correctly predicts the angular momenta in the lowest band of states for any value of $(2l, N)$ which defines the function space of the many-body system. The initial guess that the Chern–Simons gauge interaction and the Coulomb interaction between fluctuations beyond the mean field canceled is certainly not correct. The gauge field interactions are proportional to $\hbar\omega_c$ which varies linearly with $B_0$, the applied magnetic field. However, the Coulomb interactions are proportional to $e^2/\lambda$ (where $\lambda$ is the magnetic length) and vary as $B^{1/2}$. The two energy scales cannot possibly cancel for all values of $B_0$. For very large values of $B$, only the Coulomb scale is relevant in determining the low energy band of states. Our group at the University of Tennessee [18] demonstrated that the MFCF picture gave a valid description of the lowest band of states if the pair interaction energy $V(L_2)$ increased with increasing $L_2$ faster than the eigenvalue of $\hat{L}_2^2$, the square of the pair angular momentum.

Knowing this, and the occupancies of CF LLs from Jain's MFCF picture, makes it interesting to explore the correlations among the original electrons. We do this using correlation diagrams for small systems.

## 1.8 Correlation Diagrams for $N = 4$

We have already stated that Laughlin correlation can be described by drawing two CF lines between each pair $\langle i, j \rangle$. A CF line between $i$ and $j$ represents a correlation factor $z_{ij}$. The wave function $\Psi(1, 2, \ldots, N) = F\{z_{ij}\} G\{z_{ij}\}$ describing the IQL state at $\nu = {}^1/_3$ will contain $3(N - 1)$ CF lines emanating from each particle $i$. There are $(N - 1)$ CF lines associated with $F\{z_{ij}\}$, leaving $2(N - 1)$ CF lines associated with $G\{z_{ij}\}$. The correlation diagram for a Laughlin $\nu = m^{-1}$ filling factor is simple, because every pair has exactly the same correlations. For other states, like a state with $n_{QE}$ quasielectrons, the correlations are more complicated.

**Fig. 1.10** One contribution to $G$ for $(2l, N) = (8, 4)$

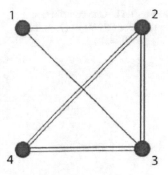

For simplicity, let's start by investigating the electron correlations for the $N = 4$ particle system with values of $2l$ in the range $9 \geq 2l \geq 6$. The values of $l_0^*$, $n_{QE}$, $l_{QE}$, $k_M$, and the total angular momentum $L$ of the lowest energy bands for these states are given in Table 1.2. We define $\kappa_F = N(N-1)/2$ as the number of CF lines appearing in the Fermi function $F\{z_{ij}\}$, and $\kappa_G$ as the number appearing in the correlation function $G\{z_{ij}\}$. Knowing $Nl$, $K_F$, and the allowed values of total angular momentum $L$, we can determine $K_G$ for each of the states listed in Table 1.2. For $l = 4.5$, 4, and 3, the corresponding values of $\kappa_G$ are 12, 8, and 6. For $l = 3.5$, there are two multiplets $L = 0$ ($\kappa_G = 8$) and $L = 2$ ($\kappa_G = 6$). We also know $k_M$ from the table. With this information, we can construct correlation functions which have to be symmetric under permutations belonging to the conjugacy class of the appropriate partition. We show one correlation diagram for each of the values of $2l$. If it is not symmetric, we must apply the appropriate symmetrization operator to the function to symmetrize it.

For $(2l, N) = (9, 4)$, there is only a single diagram; it has 2 cfs connecting each pair of particles. For a one QE state, we must partition (4) into (3, 1). The single particle $i$ belongs to subset $A$ and the other three $j, k, l$ belong to subset $B$. The latter subset has Laughlin correlations $z_{jk}^2$ between each pair belonging to $B$. Particle $i$ (in subset $A$) is the QE and has single CF lines connecting it to two of the three particles in subset $B$. Figure 1.10 shows one such diagram. The diagram corresponds to $z_{12}z_{13}z_{23}^2z_{24}^2z_{34}^2$, and this function must be appropriately symmetrized. Notice that $k_M = 5$, $Nl = 16$, and $\kappa_G = 8$, giving an $L = 2$ state for the single QE. For the two QE states with $(2l, N) = (7, 4)$, we partition (4) into (2, 2). For example, let one partition be $A = (1, 2)$ and $B = (3, 4)$. One contribution to the correlation function is shown in Fig. 1.11. This diagram corresponds to $z_{12}^2z_{14}^2z_{23}^2$, and it must be symmetrized. Notice that $k_M = 4$, $Nl = 14$, and $\kappa_G = 6$, giving $L = 2$. To obtain the $L = 0$ multiplet, we must add two more CF lines. Figure 1.12 shows one diagram for this case. It corresponds to a contribution $(z_{12}z_{23}z_{34}z_{41})^2$, and it must be symmetrized. Now $\kappa_G = 8$, and $L = 0$ results.

For $(2l, N) = (6, 4)$ we have three QEs with $k_M = 3$, and we can construct the diagram shown in Fig. 1.13. When symmetrized, it gives

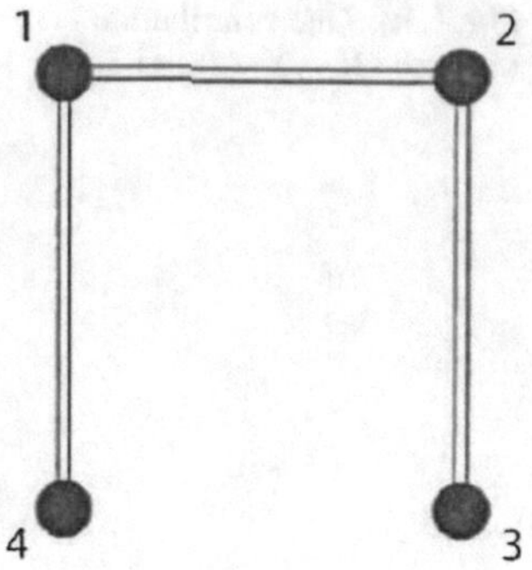

**Fig. 1.11** One contribution
to $G$ for $(2l, N) = (7, 4)$ that
gives $L = 2$

**Fig. 1.12** One contribution
to $G$ for $(2l, N) = (7, 4)$ that
gives $L = 0$

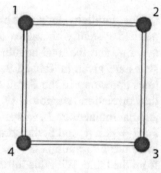

**Fig. 1.13** One contribution
to $G$ for $(2l, N) = (6, 4)$

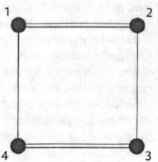

$$G\left\{z_{ij}\right\} = (z_{12}z_{34} + z_{13}z_{24})(z_{13}z_{42} + z_{14}z_{32})(z_{14}z_{23} + z_{12}z_{43}). \qquad (1.11)$$

For the four electron system, there is only one state of angular momentum $L = 0$, and
the wave function $\Psi = FG$ agrees exactly with that obtained by standard angular
momentum addition (i.e., using Clebsch–Gordan coefficients).

## 1.9 Extension to Larger Systems

For the Moore–Read state, one has $2l = 2N - 3$. We let $n = N/2$ be an integer and note that $2l = 4n - 3$ and $k_M = 2n - 2$. As discussed earlier, we partition $N$ into two equal size subsets (for example, one partition could be $A = \{1, 2, \ldots, {}^N/_2\}$ and $B = \{{}^N/_2 + 1, \ldots, N\}$). We then introduce Laughlin correlations within each subset, defining $g_A = \prod_{i<j\in A} z_{ij}^2$ and $g_B = \prod_{k<l\in B} z_{kl}^2$. Note that there are no correlations between particles in different subsets. Then the full correlation function $G$ is equal to the sum over all possible partitions of $N$ into two subsets $A$ and $B$, giving

$$G = \sum_{\text{all partitions}} g_A g_B. \tag{1.12}$$

Note that $Nl = n(2l) = n(4n - 3)$, $\kappa_F = n(2n - 1)$, and $\kappa_G = n(2n - 2)$. This yields $L = Nl - \kappa_F - \kappa_G = 0$. The requirement that $M = 0$ was necessary to describe correlations in a state with $L = 0$ was first given by Fano et al. [36].

For the Jain state at $\nu = {}^2/_5$, we know that $2l = ({}^5/_2) N - 4 = 5n - 4$ and $k_M = 3n - 3$. As with the Moore–Read state, we partition $N$ into two equal size subsets $A$ and $B$. We take Laughlin correlation within each subset giving $g_A$ and $g_B$ exactly as in the Moore–Read correlation function. Now, however, we need intersubset correlation to increase $\kappa_G$ in order to offset the increase in $Nl$ from $n(4n - 3)$ to $n(5n - 4)$ if we want to describe a state with total angular momentum $L = 0$. This gives a factor for the case $n = 3$ of

$$g_{AB} = \prod_{i\in A, j\in B} z_{ij} \hat{S}_A \left\{ (z_{16} z_{25} z_{34})^{-1} \right\}, \tag{1.13}$$

where $\hat{S}_A$ symmetrizes over all permutations belonging to the subgroup generated by the partition into the subsets $A$ and $B$. Then finally

$$G = \sum_{\text{all partitions}} g_A g_B g_{AB}.$$

We observe that $Nl = 5n - 4$, $\kappa_F = n(2n - 1)$, and $\kappa_G = n(3n - 3)$, giving $L = 0$. One (of the six) diagrams for $A = \{1, 2, 3\}$ and $B = \{4, 5, 6\}$ is shown in Fig. 1.14.

The overlap of $\Psi \propto F \{z_{ij}\} G \{z_{ij}\}$ with the exact numerical diagonalization result is almost 99%. We believe that our method should be very good for any value of $N = 2n$ in both the Moore–Read state and the Jain $\nu = {}^2/_5$ IQL state.

It is worth noting that for $N \geq 6$, there is more than one multiplet with total angular momentum $L = 0$. For example, the diagram in Fig. 1.15 has nine pairs containing Laughlin correlation factors $z_{ij}^2$, no pairs containing a single CF, and six pairs with no correlations. It is not an eigenstate of the interacting system, and it has a small overlap with the numerical diagonalization result for the IQL at $\nu = {}^2/_5$. The reason for this, even though $Nl$, $k_M$, $\kappa_F$ and $\kappa_G$ satisfy all the requirements for an $L = 0$

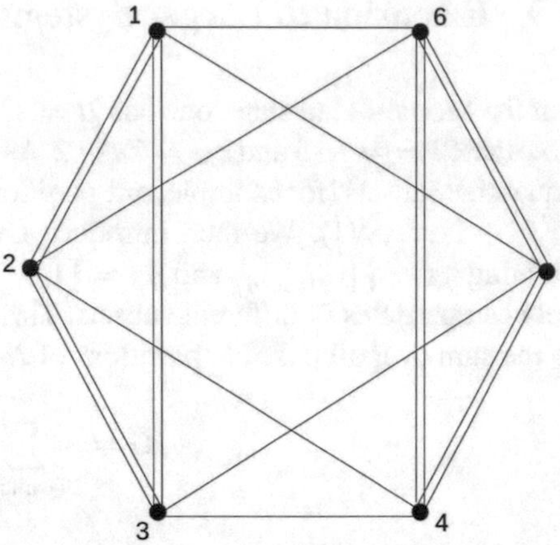

**Fig. 1.14** One correlation
diagram for the $\nu = {}^2/_5$,
$L = 0$ ground state

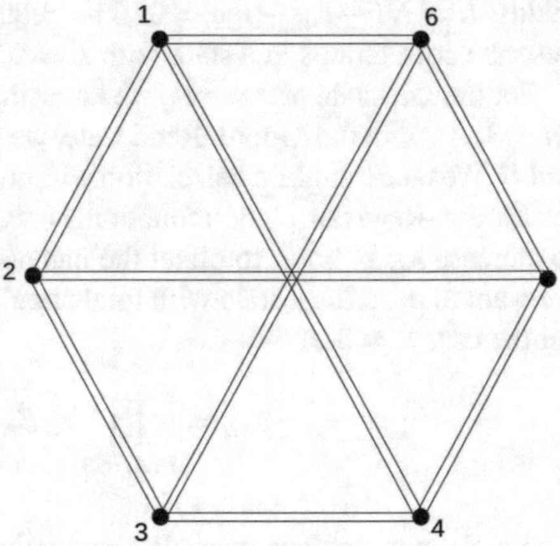

**Fig. 1.15** One diagram
containing correlations $z_{ij}^2$
among 9 of the 15 pairs

state of a $(2l, N) = (11, 6)$ system, is that the three additional Laughlin correlations
force the removal of the six pairs with single correlation factors. Because of how
$V(L_2)$ behaves with increasing $L_2$, it is not energy efficient to turn six pairs that
avoid $L_2 \leq 2l - 2$ in $G$ into three pairs that avoid $L_2 \leq 2l - 3$. We are continuing
to study other correlations, since at present we can not rigorously prove that the ones
we have selected on the basis of physical intuition are unique.

In Fig. 1.16, we show one correlation diagram containing seven pairs with Laughlin correlations and four with a single correlation factor. As in Fig. 1.15, it satisfies
all the requirements for a state with total angular momentum zero. However, the
replacement of $z_{15}z_{24}z_{26}z_{35}$ in Fig. 1.14 by $z_{14}z_{25}^2z_{36}$ in Fig. 1.16 increases the repulsive interaction energy. The correlation configurations of Figs. 1.15 and 1.16 both
could contribute to the trial wave function of an excited $L = 0$ state, but they have
essentially no overlap with the IQL ground state at $\nu = {}^2/_5$. In principle, all possible
configurations could contribute to the eigenstates with $L = 0$; however, the configuration in Fig. 1.14, or either one of the configurations in Fig. 2.4 below (they all yield

**Fig. 1.16** One diagram containing correlations $z_{ij}^2$ among 7 of the 15 pairs

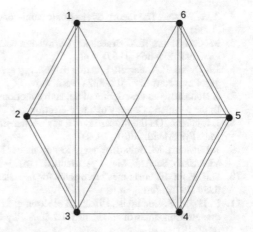

the same correlation function), gives the lowest interaction energy and has almost perfect overlap with the $L = 0$ ground state determined by numerical diagonalization. This offers valuable insight into the correlations. One can study the $L = 0$ states for different model pseudopotentials to determine how correlations change as $V(L_2)$ is varied.

It is useful to introduce the symbol $n_j$ defined as the number of pairs in a correlation diagram connected by $j$ correlation factors. Because $V(L_2)$, the interaction energy of a pair with pair angular momentum $L_2$, increases more rapidly with increasing $L_2$ than $L_2(L_2 + 1)$, the lowest energy states will be those that avoid the largest allowed values of $L_2$. For the system of $N = 6$ electrons, the $\nu = 2/5$ ground state has $(2l, N) = (11, 6)$ and $\kappa_G = 18$. Restricting $j$ to be at most 2, the allowed states can be characterized by triples $(n_2, n_1, n_0)$ with $n_2 + n_1 + n_0 = 15$ and $2n_2 + n_1 = \kappa_G = 18$. Thus, the possible states are $(9, 0, 6)$, $(7, 4, 4)$, $(6, 6, 3)$, $(4, 10, 1)$ and $(3, 12, 0)$. All diagrams corresponding to these triples must also have non-vanishing symmetrization to contribute to the eigenstate. The behaviour of the psuedopotential $V(L_2)$ for electrons in LLO yields the lowest energy state when pairs having large values of the pair angular momentum are avoided to the maximum possible extent. For a correlation diagram, the associated values of $n_j$ allow for an intuitive insight into the interaction energy.

# References

1. A. Sommerfeld, Zur elektronentheorie der metalle auf grund der fermischen statistik. Z. Phys. **47**(1–2), 1–32 (1928)
2. W. Pauli, Über den zusammenhang des abschlusses der elektronengruppen im atom mit der komplexstruktur der spektren. Z. Phys. **31**(1), 765–783 (1925)
3. F. Bloch, Über die quantenmechanik der elektronen in kristallgittern. Z. Phys. **52**(7–8), 555–600 (1929)

4. A.H. Wilson, The theory of electronic semi-conductors. Proc. R. Soc. A **133**(822), 458–491 (1931)
5. M. Gell-Mann, K.A. Brueckner, Correlation energy of an electron gas at high density. Phys. Rev. **106**, 364–368 (1957). Apr
6. J.J. Quinn, R.A. Ferrell, Electron self-energy approach to correlation in a degenerate electron gas. Phys. Rev. **112**, 812–827 (1958). Nov
7. J. Hubbard, The description of collective motions in terms of many-body perturbation theory, ii: the correlation energy of a free-electron gas. Proc. R. Soc. A **243**(1234), 336–352 (1958)
8. J. Lindhard, On the properties of a gas of charged particles. Kgl. Danske Videnskab. Selskab, Mat.-Fys. Medd. **28**(8), (1954)
9. J. Lindhard, M. Scharff, Energy loss in matter by fast particles of low charge. Kgl. Danske Videnskab. Selskab, Mat.-Fys. Medd. **27**(15), 1–31 (1953)
10. D.R. Penn, Electron mean free paths for free-electron-like materials. Phys. Rev. B **13**, 5248–5254 (1976). Jun
11. L. Hedin, S. Lundqvist, Effects of electron-electron and electron-phonon interactions on the one-electron states of solids, in *Solid State Physics*, vol. 23 (Academic Press, Dublin, 1970), pp. 1–181
12. L.D. Landau, The theory of a Fermi liquid, J. Exp. Theor. Phys. **3**, 920 (1957); Zh. Eksp. Teor. Fiz. **30**, 1058 (1956)
13. V.P. Silin, Theory of a degenerate electron liquid, J. Exp. Theor. Phys. **6**, 387 (1958); Zh. Eksp. Teor. Fiz. **33**, 495 (1957)
14. D.C. Tsui, H.L. Stormer, A.C. Gossard, Two-dimensional magnetotransport in the extreme quantum limit. Phys. Rev. Lett. **48**, 1559–1562 (1982). May
15. R.B. Laughlin, Anomalous quantum Hall effect: an incompressible quantum fluid with fractionally charged excitations. Phys. Rev. Lett. **50**, 1395–1398 (1983)
16. J.K. Jain, Composite-fermion approach for the fractional quantum Hall effect. Phys. Rev. Lett. **63**, 199–202 (1989). Jul
17. J.K. Jain, Theory of the fractional quantum Hall effect. Phys. Rev. B **41**, 7653–7665 (1990)
18. J.J. Quinn, A. Wójs, K.-S. Yi, G. Simion, The hierarchy of incompressible fractional quantum Hall states. Phys. Rep. **481**(34), 29–81 (2009)
19. X.M. Chen, J.J. Quinn, Angular momenta of composite Fermion excitations and the band structure of fractional quantum Hall systems. Solid State Commun. **92**(11), 865–868 (1994)
20. S. Gasiorowicz, *Quantum Physics* (Wiley, New York, 1974)
21. K. von Klitzing, G. Dorda, M. Pepper, New method for high-accuracy determination of the fine-structure constant based on quantized Hall resistance. Phys. Rev. Lett. **45**, 494–497 (1980)
22. K. von Klitzing, The quantized Hall effect. Rev. Mod. Phys. **58**, 519–531 (1986)
23. F.D.M. Haldane, Fractional quantization of the Hall effect: a hierarchy of incompressible quantum fluid states. Phys. Rev. Lett. **51**, 605–608 (1983)
24. F.D.M. Haldane, E.H. Rezayi, Finite-size studies of the incompressible state of the fractionally quantized Hall effect and its excitations. Phys. Rev. Lett. **54**, 237–240 (1985)
25. F. Wilczek, *Fractional Statistics and Anyon Superconductivity*, International Journal of Modern Physics (World Scientific, Singapore, 1990)
26. P. Sitko, K.-S. Yi, J.J. Quinn, Composite fermion hierarchy: condensed states of composite fermion excitations. Phys. Rev. B **56**, 12417–12421 (1997)
27. W. Pan, H.L. Störmer, D.C. Tsui, L.N. Pfeiffer, K.W. Baldwin, K.W. West, Fractional quantum Hall effect of composite fermions. Phys. Rev. Lett. **90**, 016801 (2003)
28. J.J. Quinn, A. Wójs, Composite Fermions and the fractional quantum Hall effect: essential role of the pseudopotential. Phys. E **6**(14), 1–4 (2000)
29. J.J. Quinn, A. Wójs, Composite fermions in fractional quantum Hall systems. J. Phys. Condens. Matter **12**(20), R265 (2000)
30. A. Wójs, J.J. Quinn, Quasiparticle interactions in fractional quantum Hall systems: justification of different hierarchy schemes. Phys. Rev. B **61**, 2846–2854 (2000). Jan
31. J.J. Quinn, G.E. Simion, *Electron Correlations in Strongly Interacting Systems*, Chapter 15. (World Scientific, Singapore, 2010), pp. 237–259

32. J.J. Quinn, On the absence of higher generations of incompressible daughter states of composite Fermion quasiparticles, in *Proceedings of the 18th International Conference on Recent Progress in Many-Body Theory* (Niagara Falls, New York, 2015)

33. S.-Y. Lee, V.W. Scarola, J.K. Jain, Stripe formation in the fractional quantum Hall regime. Phys. Rev. Lett. **87**, 256803 (2001)

34. S.-Y. Lee, V.W. Scarola, J.K. Jain, Structures for interacting composite fermions: stripes, bubbles, and fractional quantum Hall effect. Phys. Rev. B **66**, 085336 (2002)

35. G. Moore, N. Read, Nonabelions in the fractional quantum Hall effect. Nucl. Phys. B **360**(23), 362–396 (1991)

36. G. Fano, F. Ortolani, E. Colombo, Configuration-interaction calculations on the fractional quantum Hall effect. Phys. Rev. B **34**, 2670–2680 (1986). Aug

37. T.T. Wu, C.N. Yang, Some properties of monopole harmonics. Phys. Rev. D **16**, 1018–1021 (1977)

38. J.J. Quinn, Constructing trial wave functions for a many electron system confined to a quantum well in a strong magnetic field. Waves Random Complex Media **24**(3), 279–285 (2014)

39. A. Cappelli, L.S. Georgiev, I.T. Todorov, A unified conformal field theory description of paired quantum Hall states. Commun. Math. Phys. **205**(3), 657–689 (1999)

40. A. Cappelli, A. Georgiev, I.T. Todorov, in *Proceedings of Supersymmetries and Quantum Symmetries (SQS)*, ed. by E. Ivanov, S. Krivonov, A. Pasher (Dulna, 1999)

41. S.B. Mulay, J.J. Quinn, M.A. Shattuck, A generalized polynomial identity arising from quantum mechanics. Appl. Appl. Math. **11**, 576–584 (2016)

# Chapter 2
# Correlation Functions

## 2.1 Introduction

In this chapter, our focus is on providing a comprehensive and mathematically rigorous treatment of the correlation diagrams and their associated correlation functions. From the previous chapter, recall that a correlation diagram for $N$ (where $N$ is tacitly assumed to be at least 3) indistinguishable Fermions graphically exhibits the potencies of their mutual correlations. In purely mathematical terms, such a diagram is an undirected, loopless *multi-graph* on $N$ vertices. Here, the term *multi-graph* simply means a graph in which there may be multiple (albeit, finitely many) edges between a vertex-pair. In what follows, we regard *correlation diagram* and *multi-graph* as equivalent terms. Let $\Gamma$ be a multi-graph on $N$ vertices with some chosen labeling of its vertices by numbers $1, 2, \ldots, N$. Then, to $\Gamma$ corresponds a product of the terms $(z_i - z_j)^{p_{ij}}$, denoted by $\mu(\Gamma)$, in which $z_1, \ldots, z_N$ are indeterminates and for $1 \leq i < j \leq N$, the nonnegative integer $p_{ij}$ is the number of edges between the vertices labeled $i$ and $j$ (in $\Gamma$). In the classical theory of invariants, $\mu(\Gamma)$ is called the *graph-monomial* of $\Gamma$. Note that since our $N$ Fermions are indistinguishable, we must consider each of the possible choices of vertex-labelings, for the correlation diagram under consideration, on an equal footing. Two multi-graphs $\Gamma_1$ and $\Gamma_2$, each with $N$ labeled vertices, are said to be *isomorphic* provided one can be obtained from the other by a relabeling of its vertices (see Fig. 2.1 for an example of isomorphic multi-graphs).

The isomorphism class of a correlation diagram whose vertices are labeled by $1, \ldots, N$ is to be thought of as a *configuration* of correlated Fermions; nonisomorphic correlation diagrams correspond to distinct configurations. So, a configuration is a set of $N!$ correlation diagrams. Each correlation diagram of a given configuration has its graph-monomial. The *correlation function* of a configuration is defined to be the sum (or, if preferred, it can also be defined as the average) of the graph-monomials associated with that configuration. In other words, if we pick one correlation diagram $\Gamma$ for the configuration of $N$ Fermions and let $f(z_1, \ldots, z_N) := \mu(\Gamma)$, then the correlation function of the configuration is the *symmetrization* of $f$, i.e., $\sum f(z_{\sigma(1)}, \ldots, z_{\sigma(N)})$, where the sum ranges over all permutations $\sigma$ of $\{1, 2, \ldots, N\}$. Clearly, such a correlation function is a homogeneous polynomial symmetric in $z_1, z_2, \ldots, z_N$. If this

© Springer Nature Switzerland AG 2018
S. Mulay et al., *Strong Fermion Interactions in Fractional Quantum Hall States*, Springer Series in Solid-State Sciences 193,
https://doi.org/10.1007/978-3-030-00494-1_2

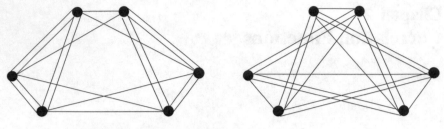

**Fig. 2.1** Isomorphic multi-graphs on 6 vertices

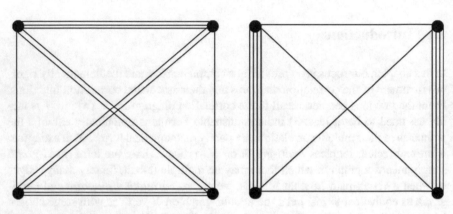

**Fig. 2.2** Equivalent configurations; IQL state, $N = 4$, $\nu = 1/3$

correlation function is identically zero, then we deem the configuration as *nonexistent*. If correlation functions of two configurations are the same up to a nonzero numerical (rational) factor, then the configurations are regarded as *equivalent*. In Fig. 2.2, the correlation function of the configuration corresponding to the diagram on the right is 2-times that of the same for the diagram on the left (the sixth example following Theorem 12 presents an even more interesting case of this type of equivalence).

It is worth noting that on account of the symmetries inherent to a given multi-graph $\Gamma$, it can very well be the case that certain distinct labelings of the vertices of $\Gamma$ yield the same graph-monomial. From a computational point of view, the correlation function of a configuration is easier to deal with when its corresponding set of graph-monomials is small and hence multi-graphs with many intrinsic symmetries are perhaps more desirable. In the extremal example of a multi-graph in which the number of edges between any two vertices is the same integer $e$ (i.e., $p_{ij} = e$ for $1 \le i < j \le N$), there are at most two distinct graph-monomials for the associated configuration; in fact, these graph-monomials differ only by a factor of $\pm 1$. Recall that such is precisely the case if we consider the Laughlin configuration for the IQL state with filling factor $\nu = 1/(2p + 1)$ (forcing $p_{ij} = e = 2p$). In general, a simple exercise in algebra shows that the graph-monomial of a correlation diagram of $N$ Fermions is a symmetric polynomial in the variables $z_1, z_2, \ldots, z_N$ if and only if there is an integer $p$ such that $p_{ij} = 2p$ for all $1 \le i < j \le N$.

For a system of $N$ correlated Fermions, the individual angular momenta of the Fermions together with the filling factor $\nu$, dictate an upper bound $d$ on the degree of a vertex (i.e., the number of edges emanating from a vertex) in any corresponding correlation diagram, whereas the total angular momentum $L$ of the system mandates that the corresponding correlation function be a homogeneous polynomial of (total) degree $(Nd/2) - L$ in $z_1, \ldots, z_N$. Generally, there are several possible configurations that meet these dictated requirements; their number increases rather steeply with increasing values of $N$. To determine which of these configurations actually exist, it is essential to ascertain the nonzero-ness of their corresponding correlation functions. This is a nontrivial task when the associated correlation diagram has vertex-pairs that are connected by an odd number of edges. Even more challenging is the problem of determining, in some concrete manner, the set of equivalence classes of these configurations. The simplest, but comparatively rare, example that can be worked out by hand is afforded by a system of 4 Fermions in an IQL state with filling factor $2/5$; in this case, each vertex of a correlation diagram must have degree 3 and then it turns out that there is only one existent configuration. Another interesting example is that of a system of 4 Fermions in an IQL state with filling factor $1/3$. In this case, each correlation diagram is a 6-regular multi-graph on 4 vertices. Of course, one of these is the (clearly existent) Laughlin configuration of Fermions corresponding to the multi-graph on 4 vertices with each pair of vertices connected by exactly 2 edges. But in addition to the Laughlin configuration, there are 6 other existing configurations as in Table 2.1.

In general, if $L = 0$ for a configuration, then it turns out that each vertex in any of its correlation diagrams must have the same maximum allowed degree $d$. An undirected, loopless multi-graph each of whose vertices has the same degree $d$, is said to be $d$-regular. The problem of counting the number of distinct configurations of $N$ Fermions with $L = 0$ and a prescribed filling factor $\nu$ translates to counting the number of isomorphism classes of $d$-regular, loopless multi-graphs on $N$ vertices. We wish to point out that at present, this counting problem appears to be largely open and it is a subject of ongoing research (see [1]). In the third section of this chapter, we establish Theorems 7–12 which help ascertain the existence of several configurations of $N$ Fermions (for arbitrary $N$). Presently, the problem of classifying configurations up to equivalence remains unsolved even for the known set of existent configurations.

For $N$ Fermions in an IQL state with filling factor $\nu = n/(2pn \pm 1) < 1/2$, we have $L = 0$ and $d = ((2p - 1)n \pm 1)((N/n) - 1)$. In particular, if $\nu = 1/3$ and hence $d = 2(N - 1)$, thanks to **gtools** [2], **MAPLE**, and **SAGE**, we can present at least a small sample of the relevant counts in Table 2.2. As already observed, the Laughlin configuration, where there are exactly $2p$ edges between each pair of vertices, is singularly distinguished among the multitude of existent configurations for the filling factor $\nu = 1/(2p + 1)$ due mainly to the following two facts:

**Table 2.1** $N = 4$ & $\nu = 1/3$; existent IQL configurations

| Configs. I | Configs. II |
|:---:|:---:|
| | |

**Table 2.2**   Apparent and existent configurations, IQL state, $\nu = 1/3$

| No. of Fermions | 2 | 3 | 4 | 5 | 6 | 7 |
|---|---|---|---|---|---|---|
| No. of apparent configurations | 1 | 1 | 7 | 37 | 2274 | 864863 |
| No. of existent configurations | 1 | 1 | 7 | 33 | 1137 | 844578 |

**Fig. 2.3**  Correlations
between two teams when
$(n, b, a) = (3, 2, 1)$

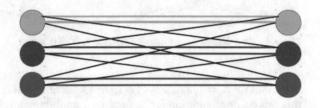

(1) the graph-monomial of its unique correlation diagram is a symmetric polynomial, and (2) the pair correlations are as small as possible. Taking a broader view
of this distinction allows us to formulate a fitting generalization of the Laughlin
configuration to the case of $N$ Fermions in an IQL state with any filling factor
$\nu = n/(2pn \pm 1) < 1/2$, where $N$ is a multiple of $n$. First, we isolate a relatively
small pool of special configurations which we feel should be the prominent contrib-
utors to the lower energy states. Imagine that the $N$ Fermions form $m := N/n$ teams,
where each team has $n$ members with distinct denominations (or ranks, or positions)
and there is no mutual correlation (repulsion) whatsoever within each team; so the
only correlations are the inter-team correlations. Furthermore, the pattern of correlations between any two teams is independent of the choice of teams, i.e., the teams
are essentially indistinguishable. Given a pair of teams, we stipulate at most two
possible types of inter-team correlations: repulsion of potency $b$ between the members of similar denominations and repulsion of potency $a$ between the members of
dissimilar denominations (which exist only when $n \geq 2$). So, the $n$ denominations
are also essentially indistinguishable. If the Fermions in such a configuration are
regrouped by denominations, then we obtain $n$ groups containing $m$ Fermions each,
where for each group, the intra-group correlations are of the same potency $b$ and
the total correlation potency between two groups is (the even integer) $m(m - 1)a$
(see remarks preceding Theorem 13). The graphic in Fig. 2.3 illustrates correlations
between two such teams, each containing three Fermions, for the $\nu = 3/7$ IQL state;
specifically, $n = 3$, $b = 2$ and $a = 1$.

Now the condition $L = 0$ translates to the requirement

$$(n - 1)a + b = (2p - 1)n \pm 1. \qquad (*)$$

As long as either $n = 1$ or $a$ and $b$ both are positive, we are assured of an existent
configuration (see Theorem 8 and its corollary). Also, if $a$, $b$ both are even integers,
then too we have an existent configuration (see Theorem 7). On the other hand, if
$a = 0$ and $n$ is even, then we have a non-existent configuration (see remarks preceding
Theorem 13). In any case, we identify this pool as the set of *balanced* configurations.

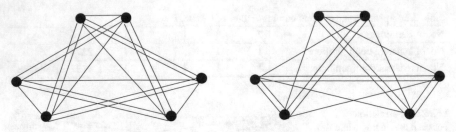

**Fig. 2.4** The minimal configurations; IQL state, $N = 6$, $\nu = 2/5$

Note that for $n = 1$, i.e., $\nu = 1/(2p + 1)$, there is just one balanced configuration; this unique configuration is the familiar Laughlin configuration. Further, we define *minimal configurations* to be those balanced configurations for which $\max\{a, b\}$ is at its least possible value. Obviously, the Laughlin configuration is the unique minimal configuration when $n = 1$. Fortunately, we find that there is a unique such 'minimal' solution to our constraint $(*)$ provided $n \geq 3$: namely,

$$(a, b) = \begin{cases} (2p - 1, 2p) & \text{if } \nu = n/(2pn + 1) \text{ and} \\ (2p - 1, 2p - 2) & \text{if } \nu = n/(2pn - 1), \end{cases}$$

and in the exceptional case of $n = 2$, mirror-symmetrically, $(2p, 2p - 1)$ and $(2p - 2, 2p - 1)$ are the only additional respective solutions. Since $a, b$ both are clearly positive in each of the cases, existence of minimal configurations is assured by our theorems. Observe that the total number of correlations between any two teams of $n$ Fermions in a minimal configuration is the even integer $(2p - 1)n^2 \pm n$. Therefore, if we regard each team as a 'super particle', then the configuration of these $m$ super particles is of Laughlin type. For 4 Fermions in the IQL state with filling factor $\nu = 2/5$, there is exactly one existent configuration and that configuration is also clearly a minimal configuration. In fact, if $N = 4$ and $\nu = 2/(4p \pm 1) < 1/2$, then the two aforementioned solutions to $(*)$ correspond to the same configuration. For 6 Fermions in the IQL state with filling factor $\nu = 2/5$, the three nonisomorphic correlation diagrams in Figs. 2.4 and 2.5 identify the only existent balanced configurations. Moreover, the two diagrams in Fig. 2.4 correspond to the only two minimal configurations; interestingly, these two minimal configurations have the same correlation function, i.e., they are equivalent configurations. In contrast, for a general $N = 2m$, it is not known whether the two minimal configurations corresponding to $\nu = 2/(4p \pm 1) < 1/2$ are equivalent.

The inbuilt symmetry of a correlation diagram for a minimal configuration implies that there are only $N!/(n!m!)$ distinct corresponding graph-monomials (see Theorem 13); thus, up to the factor of $n!m!$, the associated correlation function is the sum of these fewer graph-monomials. Given a system of $N$ correlated Fermions with a filling factor $\nu$, it is of some interest to isolate the configurations whose corresponding multi-graphs are such that each vertex-pair has an even (possibly zero) number of edges between them. There need not exist any such configuration for

**Fig. 2.5** A balanced
non-minimal configuration;
IQL state, $N = 6$, $\nu = 2/5$

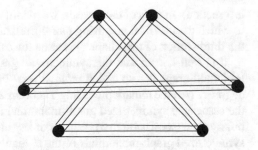

unrestricted values of $N$ and $\nu$, as demonstrated by the case of 4 Fermions in the
IQL state with filling factor $2/5$. Nevertheless, the results established in this chapter
allow explicit constructions of an array of configurations of this kind. This chapter
ends with sections that provide many detailed examples and specific applications of
our theorems. For the above defined balanced configurations, especially for the minimal configurations, the correlation-statistics in an associated correlation diagram is
useful for energy-computation (see the last section of the first chapter). For example, consider the correlation function of a minimal configuration of $N$ Fermions of
angular momentum $\ell$ in an IQL state with $\nu = n/(2pn + 1)$, where $n \geq 3$. In this system, the proportion uncorrelated Fermion-pairs is $(n - 1)/(N - 1)$, the proportion
of $(2p)$-correlated Fermion-pairs is $(N - n)/n(N - 1)$ and the remaining Fermion-
pairs are $(2p - 1)$-correlated. So, for a psuedopotential $V_0$, the energy of the system
is expected to vary directly with

$$\frac{(n - 1)}{(N - 1)} V_0(2\ell - 1) + \frac{(n - 1)(N - n)}{n(N - 1)} V_0(2\ell - 2p) + \frac{(N - n)}{n(N - 1)} V_0(2\ell - 2p - 1).$$

Note that, as $N$ grows larger, the proportion of uncorrelated pairs (i.e., correlation
factors of type 1 in the trial wave function) becomes negligible and the proportion
of $(2p - 1)$-correlated pairs (i.e., correlation factors of type $2p$ in the trial wave
function) tends to dominate.

Optimistically, generalizing from the IQL systems, one can expect to formulate
analogous definitions of 'balanced' and 'minimal' for systems containing quasielectrons. At present, there is very limited understanding of how this can be done in
full generality. Consequently, we are content to deal with systems containing $m$ QEs
in a single angular momentum shell above $N - m$ Fermions in an IQL state with the
filling factor $\nu = 1/3$. Even under these restrictions, we are able to construct existent
configurations of minimal type in only a handful of cases. Luckily, aided by a few
brute force computations, we manage to cover the ground when $N \leq 8$. Here, the
first stumbling block is the explicit computation of all possible values of the total
angular momentum $L$; for unlike in the case of an IQL state, $L$ can assume several
positive values. Of course, having determined the set of possible values of $L$, the
major difficulty still lies in finding existent configurations of the required type. We
have surmounted this difficulty for $m = 1, 2, N/2, (N + 1)/2, (N/2) + 1$ (see the

last section). In each of these cases, we identify a selection of existent configurations in which the maximal occurring pair-correlation potency (defined as the 'bound' in the third section of this chapter) is at its lowest possible value.

It is well-known that the symmetrized graph-monomial of an undirected loop-less multi-graph is a so called *relative semi-invariant* of a (generic) binary form of degree $N$ (see the remark preceding Theorem 2). If the multi-graph is $d$-regular, then the associated symmetrized graph-monomial is a *relative invariant* of the degree $N$ binary form (see Theorem 5). What is of key interest in our context is the fact that the symmetrized graph-monomials of the $d$-regular multi-graphs on $N$ vertices constitute a generating set for the vector space (over the rational numbers) formed by the relative invariants of degree $d$ and weight $Nd/2$ (see Theorem 5). So, for a system of $N$ correlated Fermions with filling factor $\nu$ and $L = 0$, the set of all correlation functions of the corresponding configurations spans the vector space of the relative invariants of weight $Nd/2$, where $d$ is the aforementioned bound dictated by the parameters of the system. For such a system, it follows that we can hope to find at least one nonzero correlation function if and only if this vector space is nonzero, and the configurations associated with our system form a single equivalence class if and only if the vector space is one-dimensional. Since a generating function for dimensions of vector spaces of relative invariants (as well as relative semi-invariants) has been known for over a century and a complete list of pairs $(N, Nd/2)$ for which the corresponding space of invariants is nonzero has been recently determined (see Theorem 6), there is a noteworthy symbiosis between the the theory of correlation functions of Fermion-configurations and the theory of invariants. The relative invariance of the correlation functions under consideration should have been at the least anticipated, if not prescribed, in view of the spherical geometry encountered in the sixth section of the first chapter. Heuristically, if our $N$ Fermions are confined to move on the surface of a sphere, then to assign complex numbers to their positions, we should first regard the sphere as a complex projective line and then project from the infinity identified by a choice of *projective* coordinates. It is then natural to require that the correlation function associated with the configuration is relatively invariant under the projective coordinate changes. More formally, let $C_N(\mathbb{C}P^1)$ denote the (standard) configuration space of $N$ indistinguishable particles on a sphere. Recall that the configuration space of $N$ indistinguishable particles on a manifold $M$ is

$$C_N(M) := \frac{M^N \setminus D_N}{S_N},$$

where $D_N$ denotes the so called *big diagonal*, i.e., the subset of ordered $N$-tuples of points of $M$ with at least two coincident points. There is a natural action of $SL_2(\mathbb{C})$, or more precisely, of $PSL_2(\mathbb{C})$, on $C_N(\mathbb{C}P^1)$ and our correlation functions $G(z_1, \ldots, z_N)$ are functions well-defined on the orbits of this action (see Theorem 3 and the last section of this chapter). An interesting related fact is that when $N \geq 6$, the space of orbits, i.e., the quotient space $C_N(\mathbb{C}P^1)/PSL_2(\mathbb{C})$, is the moduli space of hyperelliptic curves of genus $(N - 2)/2$ (there is no known relationship between existent configurations and the moduli of hyperelliptic curves). The configuration space $C_N(\mathbb{C}P^1)$ may be identified as the complement of the *discriminant hypersurface* in $\mathbb{C}P^N$ by identifying each point of $C_N(\mathbb{C}P^1)$ as a binary form of degree $N$ (e.g., see [3]). The Zariski-closure of the set of points of $C_N(\mathbb{C}P^1)$, where the correlation function of a configuration vanishes, is a hypersurface in $\mathbb{C}P^N$. The *exclusion-locus* for the corresponding configuration is then defined to be the union of this hypersurface ('configuration-exclusion') and the discriminant hypersurface ('Pauli-exclusion'). Unless $\nu = 1/(2p+1)$, the exclusion-locus of a balanced configuration of 4 or more correlated Fermions in the IQL state with filling factor $\nu < 1/2$ is strictly larger than the discriminant hypersurface.

Ever since the theory of invariants was founded, explicit construction of (semi-) invariants has been of extensive interest. Though our motivation for the explicit constructions of invariants and semi-invariants formulated in Theorems 7–12 lies in building correlation functions for systems of (strongly) correlated Fermions, these theorems can also be viewed from a purely invariant theoretic standpoint; for example, a reader familiar with the classical theory of invariants will appreciate Theorem 12 and its corollary giving a construction of an infinite family of skew-invariants (i.e., relative invariants of odd weight). Another good instance of the symbiosis between the physics of correlation functions and the theory of relative semi-invariants is the fact that for systems of $N$ correlated Fermions containing QEs, the possible values of the total angular momentum $L$ are exactly those half-integers for which there exists a nonzero relative semi-invariant of weight $(Nd^*/2) - L$ (where $d^*$ is dictated by the system under consideration; for details, see the last section of this chapter). In Appendix B, we pose some open problems that have naturally arisen from our investigation; these too can be regarded as purely invariant-theoretic.

We close this introduction with remarks on the contents of sections that follow. With the desire to make our treatment compactly self-contained, we have provided a short rigorous primer on the basic theory of invariants in the next section. This primer is designed to be broadly accessible and it is specially tailored to suit our context. For a deeper, more comprehensive treatment of the theory of invariants of binary forms, the interested reader may wish to consult either the classic [4] or the more contemporary exposition [5]. In the third section, we prove theorems that provide explicit constructions of invariants and semi-invariants. Subsequent sections present applications of these theorems to determine existent configurations. Although multi-graphs can be visually pleasing, it is undoubtedly simpler to deal with their adjacency matrices in attempting to prove precise results. Thus the reader will find our definitions and theorems formulated in the language of matrices. To meet the demands of clarity and rigor, our definitions, theorems, proofs, etc., are phrased in technical (mathematical) terms. To ease the reader's burden of navigating through these needed technicalities, plenty of explanatory remarks and illustrative concrete examples are included.

**Notation and Preliminaries**

In the rest of this chapter, $\mathbb{Z}$ denotes the set of ordinary integers, $\mathbb{N}$ denotes the set of nonnegative integers, $\mathbb{Q}$ denotes the set of rational numbers and as in the introduction, $\mathbb{C}$ denotes the set of complex numbers. Given a function $f$ defined on a set $S$, by $f(S)$, we mean the set $\{f(a) \mid a \in S\}$. Adhering to the standard practice, we let $|S|$ denote the cardinality of a set $S$. Likewise, we tacitly follow the convention that $-\infty$ is the maximum of the empty set and $\infty$ is the minimum of the empty set. The reader is assumed to be familiar with abstract algebraic notions such as *groups, rings, integral domains, fields, morphisms, characteristic, degree, order, etc.* For the basic definitions and theorems of abstract algebra needed in the following sections, any one of the standard graduate-texts on abstract algebra should suffice; our choice of a rather comprehensive and advanced reference for abstract algebra is the two-volume text [6]. All rings considered in this chapter are tacitly assumed to be commutative rings with a nonzero multiplicative identity 1 (often called 'unity'). In the next section, where definitions and theorems in the theory of invariants are presented, we do consider integral domains and fields of positive characteristic since this permits full generality from a purely mathematical point of view. The reader mainly interested from the standpoint of physics may very well suppose that all the fields encountered here contain the rational numbers (thereby assuming their characteristic to be zero).

## 2.2 Invariant-Theoretic Essentials

In this section, we only consider commutative rings with nonzero unity. As usual, $\mathrm{GL}(n, k)$ denotes the multiplicative group of $n \times n$ invertible matrices having entries in an integral domain $k$ and $\mathrm{SL}(n, k)$ denotes the subgroup of $\mathrm{GL}(n, k)$ consisting of matrices of determinant 1. Since *degree* of a rational function plays a crucial role in this section, it is helpful to briefly recall its definition and basic properties. Consider a rational function $f$ in a set of indeterminates $z$; say $f = P/Q$ for some nonzero polynomials $P$ and $Q$ in $z$ having coefficients in an integral domain $k$. Then the degree of $f$ is defined to be the difference between the (usual) degrees of $P$ and $Q$, i.e., $deg(f) = deg(P) - deg(Q)$. By convention, 0 has degree $-\infty$. Let $g$ be another rational function in $z$ with coefficients in $k$. Recall that the degree of $fg$ is the sum of the degrees of $f$ and $g$, whereas the degree of $f + g$ is bounded above by the maximum of the degrees of $f$ and $g$. Moreover, the degree of $f + g$ is the maximum of the degrees of $f$ and $g$ whenever $f$ and $g$ have unequal degrees.

**Definitions** Let $N$ be a positive integer, let $k$ be a commutative ring with $1 \neq 0$ and let $X, z_1, \ldots, z_N$ be indeterminates.

1. Given $I := (i_1, \ldots, i_N) \in \mathbb{N}^N$, let $z^I$ stand for the power-product $z_1^{i_1} \cdots z_N^{i_N}$. A polynomial $f \in k[z_1, \ldots, z_N]$ is expressed as

$$f = \sum_{I \in \mathbb{N}^N} c(I) \, z^I, \qquad \text{where } c(I) \in k \text{ for all } I \in \mathbb{N}^N.$$

   Define the *support* of $f$ to be the set

$$suppt(f) := \left\{ I \in \mathbb{N}^N \mid c(I) \neq 0 \right\}.$$

Let $l(f)$ denote the maximum of $suppt(f)$ with respect to the lexicographic ordering of $\mathbb{N}^N$ (by convention, the maximum of the empty set is $(-\infty, \ldots, -\infty)$). By the *leading coefficient* of $f$, we mean $c(l(f))$.

2. Let $\{e_1, \ldots, e_N\} \subset k[z_1, \ldots, z_N]$ denote the usual *elementary symmetric polynomials* in $\{z_1, \ldots, z_N\}$; they are defined by the equation

$$X^N + e_1 X^{N-1} + \cdots + e_N := \prod_{i=1}^{N} (X + z_i).$$

3. Assume $N$ is a unit of $k$. Define polynomials $\{y_1, \ldots, y_{N-1}\} \subset k[z_1, \ldots, z_N]$ by the equation

$$X^N + y_1 X^{N-2} + \cdots + y_{N-1} := \prod_{i=1}^{N} \left( X + z_i - \frac{(z_1 + \cdots + z_N)}{N} \right).$$

4. By $S_N$, we denote the group of all permutations of $\{1, \ldots, N\}$. Let

$$Symm_N : k[z_1, \ldots, z_N] \rightarrow k[z_1, \ldots, z_N]$$

be the ($k$-linear) *Symmetrization operator* defined by

$$Symm_N (f(z_1, \ldots, z_N)) := \sum_{\sigma \in S_N} f(z_{\sigma(1)}, \ldots, z_{\sigma(N)}).$$

A polynomial $f \in k[z_1, \ldots, z_N]$ is said to be *symmetric* provided

$$f(z_{\sigma(1)}, \ldots, z_{\sigma(N)}) = f(z_1, \ldots, z_N) \quad \text{for all } \sigma \in S_N.$$

*Remarks*

1. It is straightforward to verify that

$$k[\{z_i - z_j \mid 1 \leq i < j \leq N\}] = k[(z_1 - z_2), \ldots, (z_1 - z_N)],$$

and since $(z_1 - z_2), \ldots, (z_1 - z_N)$ are algebraically independent over $k$, the ring on the right is (naturally isomorphic to) a polynomial ring in $N - 1$ indeterminates.

2. The elementary symmetric polynomials $e_1, \ldots, e_N$ are algebraically independent over $k$. So, $k[e_1, \ldots, e_N]$ is a polynomial ring in $N$ variables.

3. For $1 \leq j \leq N$, we have

$$l(e_j) = (a_1, \ldots, a_N), \quad \text{where} \quad a_i := \begin{cases} 1 & \text{if } 1 \le i \le j, \\ 0 & \text{if } j+1 \le i \le N. \end{cases}$$

Note that the leading coefficient of $e_j$ is 1 for $1 \le j \le N$.

**Lemma 1** *For* $a := (a_1, \ldots, a_N) \in \mathbb{N}^N$, *let* $e^a$ *denote the product* $e_1^{a_1} e_2^{a_2} \cdots e_N^{a_N}$. *Given* $d := (d_1, \ldots, d_N) \in \mathbb{N}^N$, *we have* $l(e^a) = d$ *if and only if* $a_N = d_N$ *and* $a_i = d_i - d_{i+1}$ *for* $1 \le i \le N-1$. *In particular, given a* $d := (d_1, \ldots, d_N) \in \mathbb{N}^N$ *such that* $d_i \ge d_{i+1}$ *for* $1 \le i \le N-1$, *there is a unique* $a := (a_1, \ldots, a_N) \in \mathbb{N}^N$ *with* $l(e^a) = d$.

*Proof* By the last of the above remarks, $l(e^a) = (d_1, \ldots, d_N)$, where

$$d_m = \sum_{i=m}^N a_i \quad \text{for } 1 \le m \le N.$$

In particular, $d_i \ge d_{i+1}$ for $1 \le i \le N-1$. For each positive integer $r$, let $U_r$ denote the $r \times r$ lower-triangular matrix $[u_{ij}]$, where $u_{ij} := 1$ for $1 \le j \le i \le r$. Note that $U_r$ is in $SL(r, \mathbb{Z})$ for all positive integers $r$. Elements of $\mathbb{N}^r$ are regarded as $1 \times r$ (row) matrices. The above displayed equations for the components of $d$ amount to the matrix equation $d = aU_N$. Fix a $d := (d_1, \ldots, d_N) \in \mathbb{N}^N$ such that $d_1 \ge d_2 \ge \cdots \ge d_N$. By induction on $N$, we construct an $a := (a_1, \ldots, a_N) \in \mathbb{N}^N$ such that $l(e^a) = d$. More precisely, we show that

$$dU_n^{-1} = (d_1 - d_2, d_2 - d_3, \ldots, d_{n-1} - d_n, d_n)$$

for all positive integers $n$. To start with, for $n = 1$, let $a_1 := d_1$. Henceforth, assume $n \ge 2$. Observe that

$$U_n = \begin{bmatrix} U_{n-1} & 0 \\ \alpha_{n-1} & 1 \end{bmatrix} \quad \text{and} \quad U_n^{-1} = \begin{bmatrix} U_{n-1}^{-1} & 0 \\ -\alpha_{n-1} U_{n-1}^{-1} & 1 \end{bmatrix},$$

where $\alpha_{n-1}$ is the $1 \times (n-1)$ matrix with all entries 1. Letting $\delta_{n-1} := (d_1, \ldots, d_{n-1})$, we get

$$dU_n^{-1} = ((\delta_{n-1} - d_n\alpha_{n-1})U_{n-1}^{-1}, d_n).$$

By our induction hypothesis,

$$(\delta_{n-1} - d_n\alpha_{n-1})U_{n-1}^{-1} = (d_1 - d_2, d_2 - d_3, \ldots, d_{n-1} - d_n).$$

Identifying $\mathbb{N}^{n-1} \times \mathbb{N}$ with $\mathbb{N}^n$ completes the inductive step. Invertibility of $U_n$ ensures the asserted uniqueness. $\qquad\square$

**Theorem 1** *Let* $k$ *be a commutative ring with* $1 \ne 0$.

(i) *A polynomial* $\phi(z_1, \ldots, z_N)$, *with coefficients in* $k$, *is symmetric in* $z_1, \ldots, z_N$ *if and only if* $\phi(z_1, \ldots, z_N) \in k[e_1, \ldots, e_N]$.

*(ii)  Assume N is a unit of k. Then,*

$$k[e_1, y_1, \ldots, y_{N-1}] = k[e_1, \ldots, e_N].$$

*(iii)  Assume N is a unit of k. Then,*

$$k[y_1, \ldots, y_{N-1}] = k[e_1, \ldots, e_N] \cap k[(z_1 - z_2), \ldots, (z_1 - z_N)].$$

*Proof* Clearly, $k[e_1, \ldots, e_N]$ consists of symmetric polynomials. Conversely, fix a nonzero symmetric polynomial $f \in k[z_1, \ldots, z_N]$. Since $f$ is symmetric, given an $a \in suppt(f)$ and a $\theta \in S_N$, it follows that $\theta(a) \in supp(f)$. Let $d := l(f)$ and let $c(d) \in k$ denote the leading coefficient of $f$. Say $d = (d_1, \ldots, d_N)$. Since for each permutation $\theta \in S_N$, we must have $d = l(f) \geq \theta(d)$, it follows that $d_1 \geq d_2 \geq \cdots \geq d_N$. By the above Lemma 1, there is an $a \in \mathbb{N}^N$ such that $l(e^a) = d$. Then, $l(f - c(d)e^a) < d$. Moreover, $f - c(d)e^a$ is symmetric. Since there are only finitely many elements of $\mathbb{N}^N$ that are strictly less than $d$, a finite iteration of this procedure yields the zero polynomial. This proves (i).

Let $f(X) := X^N + e_1 X^{N-1} + \cdots + e_N$. Observe that $y_i$ is the coefficient of $X^{N-i-1}$ in $g(x) := f(X - e_1/N)$ for $1 \leq i \leq N - 1$, i.e., $g(X) = X^N + y_1 X^{N-2} + \cdots + y_{N-1}$. It follows that $k[e_1, y_1, \ldots, y_{N-1}] \subseteq k[e_1, \ldots, e_N]$. Since $f(X) = g(X + e_1/N)$, we infer that $k[e_1, \ldots, e_N] \subseteq k[e_1, y_1, \ldots, y_{N-1}]$. This proves (ii).

From the definition of $y_1, \ldots, y_{N-1}$, it follows that

$$g(X) = \prod_{i=1}^{N} \left( X - \frac{(z_1 - z_i) + \cdots + (z_N - z_i)}{N} \right).$$

Also, we have $z_j - z_i = (z_1 - z_i) - (z_1 - z_j)$ for $1 \leq i, j \leq N$. Then,

$$k[y_1, \ldots, y_{N-1}] \subseteq k[e_1, \ldots, e_N] \cap k[(z_1 - z_2), \ldots, (z_1 - z_N)].$$

Let $p$ be an element of the intersection appearing on the right in the above containment. In view of (ii), we have

$$p = \sum_{i=0}^{m} \lambda_i e_1^{m-i},$$

where $m$ is a nonnegative integer and $\lambda_i \in k[y_1, \ldots, y_{N-1}]$ for $0 \leq i \leq m$. Let $t$ be an indeterminate and substitute $z_i + t$ for each $z_i$ in the above equality. Since $p$ and the $\lambda_j$ are in $k[(z_1 - z_2), \ldots, (z_1 - z_N)]$, our substitution leaves them unchanged, whereas $e_1$ changes to $e_1 + Nt$ and $N \neq 0$ in $k$. From the algebraic independence of $t, e_1, y_1, \ldots, y_{N-1}$ over $k$, it follows that $m = 0$, i.e., $p$ is in $k[y_1, \ldots, y_{N-1}]$. This proves (iii). $\qquad\square$

**Corollary** *Let $k$ be an integral domain such that $N$ is a unit of $k$. For a polynomial $f \in k[z_1, \ldots, z_N]$, consider the following.*

*(i) $f \in k[y_1, \ldots, y_{N-1}]$.*

*(ii) $f$ is symmetric in $z_1, \ldots, z_N$ and $f(z_1 + c, \ldots, z_N + c) = f(z_1, \ldots, z_N)$ for all $c \in k$.*

*(iii) $f$ is symmetric in $z_1, \ldots, z_N$ and $f(z_1 + t, \ldots, z_N + t) = f(z_1, \ldots, z_N)$ for an indeterminate $t$.*

*Then, (i) $\Rightarrow$ (ii), (iii) $\Rightarrow$ (i) and if $k$ is infinite, then (ii) $\Rightarrow$ (iii). In particular, if $k$ is infinite, then (i), (ii) and (iii) are equivalent.*

*Proof* It is straightforward to see that (i) implies (ii) in view of assertion (iii) of Theorem 1. Now assume $k$ is infinite and (ii) holds. Write

$$ f(z_1 + t, \ldots, z_N + t) - f(z_1, \ldots, z_N) =: g_0 t^d + \cdots + g_d - f, $$

with $g_i \in k[z_1, \ldots, z_N]$ for $0 \le i \le d$. By assumption, the substitution $t = c$, where $c \in k$, makes the left side of the above equation 0. Hence the polynomial on the right, which is a polynomial in $t$ with coefficients in an integral domain $k[z_1, \ldots, z_N]$, has $k$ as a subset of its roots. Since a nonzero polynomial in $t$ having coefficients in an integral domain can have only finitely many roots in that integral domain, the infinitude of $k$ forces the polynomial on the right to be identically zero. Thus (ii) implies (iii). Finally, letting $u_j := z_1 - z_j$ for $2 \le j \le N$, we have $k[z_1, \ldots, z_N] = k[z_1, u_2, \ldots, u_N]$. Now let

$$ f(z_1, \ldots, z_N) = f(z_1, z_1 - u_2, \ldots, z_1 - u_N) =: h_0 z_1^r + \cdots + h_r, $$

where $h_i$ are in $k[u_2, \ldots, u_N]$. Since each member of $k[u_2, \ldots, u_N]$ is invariant under the translation of the $z_i$ by $t$, assumption (iii) leads to the equation

$$ h_0 (z_1 + t)^r + \cdots + h_r = h_0 z_1^r + \cdots + h_r. $$

Comparing the $t$-degree of both sides, we conclude that $r = 0$, i.e., $f = h_0 \in k[u_2, \ldots, u_N]$. Moreover, $f$ being symmetric, assertion (iii) of Theorem 1 assures that $f \in k[y_1, \ldots, y_{N-1}]$. Thus (iii) implies (i). $\qquad \square$

**Definitions** Let $k$ be a commutative ring with $1 \ne 0$ and let $e_0$ be an indeterminate.

1. Given a polynomial $f \in k[e_0, e_1, \ldots, e_N]$, where

$$ f := \sum c_{m_1 m_2 \ldots m_N} \left( \prod_{i=0}^{N} e_i^{m_i} \right) $$

with $c_{m_1 m_2 \ldots m_N} \in k$, define its *e-weighted degree* to be

$$\max \left\{ \sum_{i=1}^{N} i\, m_i \;\middle|\; c_{m_1 m_2 \ldots m_N} \neq 0 \right\}.$$

$f$ is said to be *e-weighted homogeneous* of $e$-weighted degree $d$ if

$$\sum_{i=1}^{N} i\, m_i = d \qquad \text{whenever } c_{m_1 m_2 \ldots m_N} \neq 0.$$

2. Assume $N$ is a unit of $k$. Given a polynomial $f \in k[y_1, \ldots, y_{N-1}]$, where

$$f := \sum c_{m_1 m_2 \ldots m_{N-1}} \left( \prod_{i=1}^{N-1} y_i^{m_i} \right)$$

with $c_{m_1 m_2 \ldots m_N} \in k$, define its *y-weighted degree* to be

$$\max \left\{ \sum_{i=1}^{N-1} (i+1)\, m_i \;\middle|\; c_{m_1 m_2 \ldots m_{N-1}} \neq 0 \right\}.$$

$f$ is said to be *y-weighted homogeneous* of $y$-weighted degree $d$ if

$$\sum_{i=1}^{N-1} (i+1)\, m_i = d \qquad \text{whenever } c_{m_1 m_2 \ldots m_{N-1}} \neq 0.$$

3. For integers $m$ and $d$, let $W(m, d, N)$ be the set of $f \in k[e_0, e_1, \ldots, e_N]$ such that $f$ is $e$-weighted homogeneous of $e$-weighted degree $m$ and $f$ is homogeneous in $e_0, \ldots, e_N$ of degree $d$. Also, let $V(m, d, N)$ be the set of $v \in k[e_1, \ldots, e_N]$ such that $v$ is $e$-weighted homogeneous of $e$-weighted degree $m$ and the total degree of $v$ in $e_1, \ldots, e_N$ is at most $d$.

4. Assume $N$ is a unit of $k$. Let $\partial$ and $\mathfrak{D}$ be the $k$-derivations of $k[e_0, e_1, \ldots, e_N]$ such that $\partial e_1 = N e_0$, $\partial e_0 = 0$, $\partial y_i = 0$ for $1 \leq i \leq N - 1$, $\mathfrak{D} e_N = 0$ and $\mathfrak{D} e_i = (i+1) e_{i+1}$ for $0 \leq i \leq N - 1$. Let $\partial_1$ be the $k$-derivation of $k[e_1, y_1, \ldots, y_{N-1}]$ such that $\partial_1 e_1 = N$ and $\partial_1 y_i = 0$ for $1 \leq i \leq N - 1$.

5. Assume $k$ is a field and $N$ is a unit of $k$. For nonnegative integers $m$ and $d$, define $\mathcal{H}_k(m, N)$ to be the subset of $k[y_1, \ldots, y_{N-1}]$ consisting of polynomials that are $y$-weighted homogeneous of $y$-weighted degree $m$. Elements of $\mathcal{H}_k(m, N)$ are called *semi-invariants* of weight $m$ of the *binary form* $X^N + e_1 X^{N-1} Y + \cdots + e_j X^{N-j} Y^j + \cdots + e_N Y^N$.

6. Let $\mathcal{H}_k(m, d, N) := \{ h \in \mathcal{H}_k(m, N) \mid h \text{ has total degree } \leq d \text{ in } e_1, \ldots, e_N \}$.

7. For integers $m, d$ and a positive integer $n$, define $p(m, d, n)$ to be the cardinality of the (finite) set

$$P(m, d, n) := \left\{ (a_1, \ldots, a_n) \in \mathbb{N}^n \;\middle|\; \sum_{r=1}^{n} a_r \leq d, \quad \sum_{r=1}^{n} r a_r = m \right\}.$$

*Remarks*

1. The zero polynomial is regarded to be $e$-weighted (resp. $y$-weighted) homogeneous of every $e$-weighted (resp. $y$-weighted) degree.
2. Clearly, $W(m, d, N)$ and $V(m, d, N)$ are $k$-subspaces of $k[e_0, e_1, \ldots, e_N]$. Furthermore, $V(m, d, N)$ is a $k$-subspace of $k[e_1, \ldots, e_N]$ of dimension $p(m, d, N)$.
3. The set $W(m, d, N)$ consists of degree $d$ homogenizations of the polynomials in $V(m, d, N)$ with respect to $e_0$, i.e.,

$$W(m, d, N) = \left\{ e_0^d f \left( \frac{e_1}{e_0}, \ldots, \frac{e_N}{e_0} \right) \;\middle|\; f \in V(m, d, N) \right\}.$$

4. Note that $\mathcal{H}_k(m, N)$ is a $k$-linear subspace of $k[y_1, \ldots, y_{N-1}]$ and $\mathcal{H}_k(m, d, N)$ is a $k$-subspace of $\mathcal{H}_k(m, N)$. Note that $\mathcal{H}_k(1, N) = 0$. From the Corollary to Theorem 1, it follows that if $m \geq 1$, then $\mathcal{H}_k(m, 1, N) = 0$.
5. Assume $N \geq 3$. Then, for each integer $m \geq 2$, we can find nonnegative integers $i, j$ such that $2i + 3j = m$ and hence $y_1^i y_2^j$ is in $\mathcal{H}_k(m, N)$; so, $\mathcal{H}_k(m, N) \neq 0$ for all $m \geq 2$.
6. Given a member of $\mathcal{H}_k(m, N)$, its total degree in $e_1, \ldots, e_N$ can not be readily detected, e.g., the polynomial $6y_1^3 + 40y_1 y_3 + 9y_2^2 \in \mathcal{H}_k(6, 4)$ has total degree 4 in $e_1, e_2, e_3, e_4$. Thus, in general, it is difficult to determine whether a particular member of $\mathcal{H}_k(m, d, N)$ is nonzero.
7. Obviously, $p(m, d, n) = 0$ if either $m < 0$ or $d < 0$ or $m > nd$. Also, $p(0, d, n) = 1$ if $d \geq 0$. Note that $p(m, d, n)$ is the number of partitions of $m$ into at most $d$ parts with each part at most $n$.

**Theorem 2** *Let $k$ be a commutative ring with $1 \neq 0$.*

(i) *Let $f$ be a nonzero element of $k[e_1, \ldots, e_N]$. The $e$-weighted degree of $f$ is $d$ if and only if the total degree of $f$ in $z_1, \ldots, z_N$ is $d$. Moreover, $f$ is $e$-weighted homogeneous of $e$-weighted degree $d$ if and only if $f$ is homogeneous of degree $d$ in $z_1, \ldots, z_N$.*

(ii) *Assume $N$ is a unit of $k$. Let $f$ be a nonzero element of $k[y_1, \ldots, y_{N-1}]$. The $y$-weighted degree of $f$ is $d$ if and only if the total degree of $f$ in $z_1, \ldots, z_N$ is $d$. Moreover, $f$ is $y$-weighted homogeneous of $y$-weighted degree $d$ if and only if $f$ is homogeneous of degree $d$ in $z_1, \ldots, z_N$.*

(iii) *Let $f$ be a nonzero element of $k[e_1, \ldots, e_N]$. The total degree of $f$ in $e_1, \ldots, e_N$ is $d$ if and only if the $z_i$-degree of $f$ is $d$ for $1 \leq i \leq N$.*

(iv) *Assume that $k$ is a field. Suppose $0 \neq f \in k[z_1 - z_2, \ldots, z_1 - z_N]$ is homogeneous of total degree $m$ in $z_1, \ldots, z_N$. If the $z_i$-degree of $f$ is $d_i$ for $1 \leq i \leq N$, then $2m \leq d_1 + \cdots + d_N$.*

(v) *Assume that $k$ is a field containing $\mathbb{Q}$ and $m, d$ are nonnegative integers such that $2m \leq dN + 1$. Then $\partial$ maps $W(m, d, N)$ onto $W(m - 1, d, N)$.*

*(vi) Assume that k is a field containing $\mathbb{Q}$ and m, d are nonnegative integers. If $2m > dN$, then $\mathcal{H}_k(m, d, N) = 0$. If $2m \leq dN$, then, as a vector space over k, $\mathcal{H}_k(m, d, N)$ has dimension $p(m, d, N) - p(m - 1, d, N)$.*

*Proof* Since each $e_i$ is a homogeneous polynomial of degree $i$ in $z_1, \ldots, z_N$, any power-product of $e_1, \ldots, e_N$ is also a homogeneous polynomial in $z_1, \ldots, z_N$. Now for $f$ as in (i), first assume that $f$ is $e$-weighted homogeneous of $e$-weighted degree $d$. Then $f$ is clearly homogeneous of degree $d$ in $z_1, \ldots, z_N$. Moreover, since $k[e_1, \ldots, e_N]$ is $k$-isomorphic to $k[z_1, \ldots, z_N]$, $f$ is also a nonzero polynomial in $z_1, \ldots, z_N$. Conversely, if $f$ is homogeneous of degree $d$ in $z_1, \ldots, z_N$, then its $e$-weighted degree is clearly $d$ and furthermore, $f$ must be $e$-weighted homogeneous. The rest of the assertion (i) follows easily.

Next, let $f$ be as in (ii). Since $y_i$ is homogeneous of degree $i + 1$ in $z_1, \ldots, z_N$ for $1 \leq i \leq N - 1$, our assertion follows by an argument entirely similar to the argument in the proof of (i).

To prove (iii), let $f$ be as in (iii) and first suppose that the total degree of $f$ in $e_1, \ldots, e_N$ is $d$. Since each $e_i$ is of degree 1 in $z_N$, the $z_N$-degree of $f$ can not exceed $d$. Thus it suffices to prove (iii) by assuming $f$ to be homogeneous in $e_1, \ldots, e_N$. Let $f = g(z_1, \ldots, z_N)$. Let $u_0 = 1$, $u_N = 0$ and $u_1, \ldots, u_{N-1}$ denote the elementary symmetric functions of $z_1, \ldots, z_{N-1}$, i.e.,

$$X^{N-1} + \sum_{t=1}^{N-1} u_i X^{N-1-i} = \prod_{i=1}^{N-1} (X + z_i).$$

Assume that the $z_N$-degree of ($f$ or equivalently, of) $g$ is strictly less than $d$. Let $t$ be an indeterminate and let

$$h(t) := t^d g\left(z_1, \ldots, z_{N-1}, \frac{z_N}{t}\right).$$

Then we have $h(0) = 0$. Now observe that

$$t\, e_i\left(z_1, \ldots, z_{N-1}, \frac{z_N}{t}\right) = z_N u_{i-1} + t u_i$$

for $1 \leq i \leq N - 1$ and $te_N(z_1, \ldots, z_{N-1}, z_N/t) = z_N u_{N-1}$. Hence

$$h(t) = f(z_N + tu_1, \ldots, z_N u_{N-2} + tu_{N-1}, z_N u_{N-1}).$$

But then, by substituting $t = 0$, we get the equation

$$0 = f(z_N, \ldots, z_N u_{N-2}, z_N u_{N-1}) = z_N^d f(1, u_1, \ldots, u_{N-1}).$$

This is absurd since $f$ is nonzero and $u_1, \ldots, u_{N-1}$ are algebraically independent over $k$. In conclusion, the $z_N$-degree of $f$ is $d$. Using the fact that $f$ is symmetric in $z_1, \ldots, z_N$, we conclude that the $z_i$-degree of $f$ is $d$ for $1 \leq i \leq N$. By what has been shown, if the $z_i$-degree of $f$ is $d$ for $1 \leq i \leq N$, then the total degree of $f$ in $e_1, \ldots, e_N$ can neither exceed $d$ nor be strictly less than $d$. Thus (iii) is fully established.

To prove (iv), we rely on the fact that the polynomial ring $k[z_1 - z_2, \ldots, z_1 - z_N]$ is a unique factorization domain. Since $f$ is nonzero and $z_1 - z_N$ is irreducible, there is a unique nonnegative integer $s$ such that $f = (z_1 - z_N)^s g$, where $g \in k[z_1 - z_2, \ldots, z_1 - z_N]$ is relatively prime to $z_1 - z_N$. In particular, $g = h_0 + \sum_{i \geq 1} h_i(z_1 - z_N)^i$, where $h_i \in k[z_1 - z_2, \ldots, z_1 - z_{N-1}]$ for $i \geq 0$ and $h_0 \neq 0$. Moreover, $g$ is homogeneous of total degree $m - s$ in $z_1, \ldots, z_N$, the $z_1$-degree of $g$ is $d_1 - s$, the $z_N$-degree of $g$ is $d_N - s$ and the $z_i$-degree of $g$ is $d_i$ for $2 \leq i \leq N - 1$. It follows that $h_i$ is homogeneous of total degree $m - s - i$ in $z_1, \ldots, z_{N-1}$ for $i \geq 0$. The rest of the proof proceeds by induction on $N$. If $N = 2$, then $s = m$ and $d_1 = s = d_2$; so, $2m \leq d_1 + d_2$. Henceforth, assume $N \geq 3$ and let $r_i$ denote the $z_i$-degree of $h_0$ for $1 \leq i \leq N - 1$. Clearly, $r_1 \leq d_1 + d_N - 2s$ and $r_i \leq d_i$ for $2 \leq i \leq N - 1$. By the induction hypothesis, $2(m - s) \leq r_1 + \cdots + r_{N-1}$. Hence $2m - 2s \leq d_1 + d_N - 2s + d_2 + \cdots + d_{N-1}$, i.e., $2m \leq d_1 + \cdots + d_N$ as asserted. Thus (iv) holds.

With the hypotheses and notation of (v), first observe that for $1 \leq r \leq N$,

$$e_r = \frac{1}{N^r}\binom{N}{r}e_1^r + \sum_{j=1}^{r-1} \frac{1}{N^{r-j-1}}\binom{N-j-1}{r-j-1}e_1^{r-j-1}y_j$$

and hence $\partial e_r = (N - r + 1)e_{r-1}$. It is now readily verified that for $h \in W(m, d, N)$, we have $\partial h \in W(m - 1, d, N)$. So, $\partial$ restricts to a $k$-linear transformation from $W(m, d, N)$ to $W(m - 1, d, N)$. In particular, $\partial^{m+1}h = 0$ for all $h \in W(m, d, N)$. Likewise, $\mathfrak{D}$ restricts to a $k$-linear transformation from $W(m - 1, d, N)$ to $W(m, d, N)$. Thus, for each positive integer $s$, the compositions $\partial^s \mathfrak{D}^s$, $\mathfrak{D}^s \partial^s$ are $k$-linear operators on $W(m, d, N)$. Moreover, given a nonnegative integer $i$, it is readily verified that

$$(\partial \mathfrak{D} - \mathfrak{D}\partial)h = (Nd - 2m + 2i)h \quad \text{for all } h \in W(m - i, d, N),$$

i.e., $\partial \mathfrak{D} - \mathfrak{D}\partial = (Nd - 2m + 2i)I_{m-i}$, where $I_{m-i}$ denotes the identity operator on $W(m - i, d, N)$. Likewise, for any positive integer $s$, we clearly have the following equality of $k$-linear operators on $k[e_0, e_1, \ldots, e_N]$:

$$\partial \mathfrak{D}^s - \mathfrak{D}^s \partial = \sum_{i=1}^{s} \mathfrak{D}^{i-1}(\partial \mathfrak{D} - \mathfrak{D}\partial)\mathfrak{D}^{s-i}.$$

Hence for $\alpha \in W(m - s, d, N)$,

$$\partial \mathfrak{D}^s \alpha - \mathfrak{D}^s \partial \alpha = \sum_{i=1}^{s} \mathfrak{D}^{i-1}(Nd - 2m + 2i)I_{m-i}\mathfrak{D}^{s-i}\alpha = s(Nd - 2m + s + 1)\mathfrak{D}^{s-1}\alpha.$$

Given $h \in W(m - 1, d, N)$, let $h_s := D^{s-1} \partial^{s-1} h$. Then, note that $h_1 = h$, $\mathfrak{D} h_s \in W(m, d, N)$ for all $s$ and $h_s = 0$ for all $s \geq m + 1$. Substituting $\alpha = \partial^{s-1} h$ in the above displayed identity leads to the system of equations:

$$\partial \mathfrak{D} h_s - h_{s+1} = s(Nd - 2m + s + 1)h_s \quad \text{for all } s \geq 1.$$

Since $h \in W(m - 1, d, N)$ and $2m \leq Nd + 1$, we have $h_{m+1} = 0$ and most importantly, $(Nd + s + 1 - 2m) \geq 1$ for all $s \geq 1$. So, starting from the $m$th equation in the above system and substituting backwards, we get

$$h = \partial h^*, \quad \text{where } h^* := \sum_{j=1}^{m} \frac{(-1)^{j-1}}{j! \prod_{i=1}^{j}(Nd - 2m + i + 1)} \mathfrak{D} h_j \in W(m, d, N).$$

This proves (v).

Lastly, consider the space $\mathcal{H}_k(m, d, N)$, where $m, d$ are nonnegative integers. Suppose $0 \neq h \in \mathcal{H}_k(m, d, N)$ and let $d_i$ denote the $z_i$-degree of $h$ for $1 \leq i \leq N$. Then, by (iii), we have $d_i \leq d$ for $1 \leq i \leq N$ and hence (iv) implies that $2m \leq d_1 + \cdots + d_N \leq Nd$. Consequently, $\mathcal{H}_k(m, d, N) = 0$ for pairs $(m, d)$ with $2m > Nd$. Let the *kernel* of $\partial_1$ be

$$Ker(\partial_1) := \{v \in k[e_1, y_1, \ldots, y_{N-1}] \mid \partial(v) = 0\}.$$

Since $k$ is assumed to have characteristic zero, it is clear that $Ker(\partial_1) = k[y_1, \ldots, y_{N-1}]$. Suppose $2m \leq Nd + 1$ and let $v \in V(m - 1, d, N)$. Let $h \in W(m - 1, d, N)$ be such that $v = h(1, e_1, \ldots, e_N)$. By (v), there is an $h^* \in W(m, d, N)$ with $\partial h^* = h$. Observe that for $v^* := h^*(1, e_1, \ldots, e_N)$, we have $v^* \in V(m, d, N)$ and $\partial_1 v^* = v$. In other words, if $2m \leq Nd + 1$, then $\partial_1$ restricts to a surjective $k$-linear transformation from $V(m, d, N)$ onto $V(m - 1, d, N)$. Next, for $1 \leq r \leq N - 1$, we have

$$y_r = \frac{(-1)^{r+1}}{N^{r+1}} \binom{N}{r+1} e_1^{r+1} + \sum_{j=1}^{r+1} \frac{(-1)^{r+1-j}}{N^{r+1-j}} \binom{N-j}{r+1-j} e_1^{r+1-j} e_j.$$

So, $y_r$ is $e$-weighted homogeneous of $e$-weighted degree $r + 1$ for $1 \leq r \leq N - 1$. Each $v \in V(m, d, N)$ has a unique expression $v = u_0 + \cdots + u_i e_1^i + \cdots$, where $u_i \in k[y_1, \ldots, y_{N-1}]$ for all $i \in \mathbb{N}$. For $1 \leq j \leq N - 1$, $y_j$ is $e$-weighted homogeneous as well as $y$-weighted homogeneous of the same weighted degree $j + 1$. This allows us to infer that $u_i$ is $y$-weighted homogeneous of $y$-weighted degree $m - i$ for all $i \in \mathbb{N}$. In particular, $v \in k[y_1, \ldots, y_{N-1}]$ if and only if $v = u_0 \in \mathcal{H}_k(m, d, N)$, i.e., $V(m, d, N) \cap Ker(\partial_1) = \mathcal{H}_k(m, d, N)$. So $\partial_1$ induces a $k$-linear isomorphism

$$\frac{V(m, d, N)}{\mathcal{H}_k(m, d, N)} \cong V(m - 1, d, N).$$

Thus, $\mathcal{H}_k(m, d, N)$ has dimension $p(m, d, N) - p(m - 1, d, N)$.  $\square$

*Remarks*

1. Suppose $0 \neq h \in \mathcal{H}_k(m, d, N)$ and $2m = Nd$. Then, in view of (iii) and (iv) of the above theorem, the total degree of $h$ in $e_1, \ldots, e_N$ must be $d$.
2. It is easy to verify that if $h := e_N^d \in W(Nd, d, N)$, then there does not exist $h^* \in W(Nd + 1, d, N)$ such that $\partial h^* = h$.
3. Assume $k$ is a field containing $\mathbb{Q}$. We may have $\mathcal{H}_k(m, d, N) = 0$ even when $2m \leq Nd$, e.g., $\mathcal{H}_k(5, 3)$ is easily seen to have $k$-basis $\{y_1 y_2\}$ and in this case, since $y_1 y_2$ has total degree 5 in $e_1, e_2, e_3$, we have $\mathcal{H}_k(5, 4, 3) = 0$. Note that $p(5, 4, 3) = p(4, 4, 3)$.
4. From (vi) of the above theorem, we infer that if $2m \leq Nd + 1$, then $p(m, d, N) \geq p(m - 1, d, N)$ and $p(m, d, N) \leq p(m - 1, d, N)$ if $2m \geq Nd + 2$. Since $\mathcal{H}_k((Nd + 1)/2, d, N) = 0$ by (vi), we have $p((Nd + 1)/2, d, N) = p((Nd - 1)/2, d, N)$.

**Definitions**  Let $k$ be a field.

1. Let $s$, $t$, $u$, $v$ be indeterminates and define polynomials $A_0, \ldots, A_N$ in $k[e_1, \ldots, e_N, s, t, u, v]$ by

$$\sum_{i=0}^{N} A_i X^{N-i} := (sX + t)^N + \sum_{j=1}^{N} e_j (sX + t)^{N-j} (uX + v)^j.$$

2. Define rational functions $E_1, \ldots, E_N$ by setting

$$E_j := \frac{A_j}{A_0} \quad \text{for } 1 \leq j \leq N.$$

3. By a *binary invariant of type* $(N, d)$ (and of weight $w$), we mean a polynomial $f \in k[e_1, \ldots, e_N]$ which is either 0 or of total degree $d$ in $e_1, \ldots, e_N$ and such that for some nonnegative integer $w$,

$$A_0^d f(E_1, \ldots, E_N) = (sv - tu)^w f(e_1, \ldots, e_N).$$

4. A polynomial $g \in k[z_1, \ldots, z_N]$ is called a binary invariant of type $(N, d)$ provided $g \in k[e_1, \ldots, e_N]$ and (as such) it is a binary invariant of type $(N, d)$.
5. Let $\alpha$ be an indeterminate. Given an $M := [m_{ij}] \in \mathrm{SL}(2, k)$, define

$$M(\alpha) := \frac{m_{11}\alpha + m_{12}}{m_{21}\alpha + m_{22}}.$$

6. A polynomial $f \in k[z_1, \ldots, z_N]$ is said to be a *relative* $\mathrm{SL}(2, k)$-*invariant of type* $(N, d)$ if $f$ is symmetric in $z_1, \ldots, z_N$ and

$$f(M(z_1), \ldots, M(z_N)) = \left\{ \prod_{r=1}^{N} \frac{1}{(m_{21} z_r + m_{22})^d} \right\} f(z_1, \ldots, z_N)$$

for all $M := [m_{ij}] \in SL(2, k)$.

*Remarks*

1. If $g$ is a nonzero binary invariant of type $(N, d)$, then comparing the total degrees in $s, t, u, v$ of both sides of the equation

$$A_0^d \phi(E_1, \ldots, E_N) = (sv - tu)^w \cdot g,$$

we see that $2w = Nd$.
2. In particular, if $Nd$ is odd, then there does not exist a nonzero binary invariant of type $(N, d)$.

**Theorem 3** *Let $k$ be a field and let $g \in k[z_1, \ldots, z_N]$ be a nonzero polynomial. Let*

$$f(X) := (X + z_1)(X + z_2) \cdots (X + z_N) = X^N + e_1 X^{N-1} + \cdots + e_N.$$

*Consider the following properties.*

 (i) *$g$ is a binary invariant of type $(N, d)$.*
 (ii) *$g$ is a relative $SL(2, k)$-invariant of type $(N, d)$.*
 (iii) *$g$ is symmetric in $z_1, \ldots, z_N$ and for an indeterminate $t$, we have*

$$(1) \quad g(t^2 z_1, \ldots, t^2 z_N) = t^{Nd} g(z_1, \ldots, z_N),$$
$$(2) \quad g(z_1 + t, \ldots, z_N + t) = g(z_1, \ldots, z_N), \quad and$$
$$(3) \quad g\left(\frac{t z_1}{z_1 + t}, \ldots, \frac{t z_N}{z_N + t}\right) = \left(\frac{t^N}{f(t)}\right)^d g(z_1, \ldots, z_N).$$

*Then $(i) \Rightarrow (ii)$ and $(iii) \Rightarrow (i)$; if $k$ is infinite, $(ii) \Rightarrow (iii)$. In particular, if $k$ is infinite, then assertions $(i), (ii), (iii)$ are equivalent.*

*Proof* Assume $g$ is a binary invariant of type $(N, d)$. Say $g = \phi(e_1, \ldots, e_N) \in k[e_1, \ldots, e_N]$. As before, let $s, t, u, v$ be indeterminates. Observe that

$$(sX + t)^N + \sum_{j=1}^{N} e_j (sX + t)^{N-j} (uX + v)^j = \prod_{i=1}^{N} (uz_i + s) \prod_{i=1}^{N} \left(X + \frac{v z_i + t}{u z_i + s}\right).$$

It at once follows that $A_0$ is the first product appearing on the right, i.e., $A_0 = u^N f(s/u)$ and

$$E_j = \frac{A_j}{A_0} = e_j \left(\frac{v z_1 + t}{u z_1 + s}, \ldots, \frac{v z_N + t}{u z_N + s}\right)$$

for $1 \leq j \leq N$. Thus, there is a nonnegative integer $w$ such that

$$g\left(\frac{v z_1 + t}{u z_1 + s}, \ldots, \frac{v z_N + t}{u z_N + s}\right) = \phi(E_1, \ldots, E_N) = \frac{(sv - tu)^w}{A_0^d} g(z_1, \ldots, z_N).$$

Given an $M := [m_{ij}]$ in $SL(2, k)$, substitute $v = m_{11}, t = m_{12}, u = m_{21}$ and $s = m_{22}$ in the above equation and use the fact that $M$ has determinant 1. Then $g$ is seen to be a relative $SL(2, k)$-invariant of type $(N, d)$. So (ii) follows from (i).

Now assume (ii) and assume $k$ to be an infinite field. Observe that for any $c \in k$, we have

$$g\left(\frac{z_1 + c}{0z_1 + 1}, \ldots, \frac{z_N + c}{0z_N + 1}\right) = g(z_1 + c, \ldots, z_N + c) = g(z_1, \ldots, z_N).$$

From the Corollary to Theorem 1, it follows that

$$g(z_1 + t, \ldots, z_N + t) = g(z_1, \ldots, z_N)$$

for an indeterminate $t$. Likewise, for any $0 \neq b \in k$, we have

$$g\left(\frac{bz_1 + 0}{b^{-1}}, \ldots, \frac{bz_N + 0}{b^{-1}}\right) = g(b^2 z_1, \ldots, b^2 z_N) = b^{Nd} g(z_1, \ldots, z_N);$$

also, letting $a := 1/b$, we have

$$g\left(\frac{az_1}{z_1 + a}, \ldots, \frac{az_N}{z_N + a}\right) = \left\{\prod_{r=1}^{N} \frac{a^d}{(z_r + a)^d}\right\} g(z_1, \ldots, z_N).$$

Now the polynomial $g(t^2 z_1, \ldots, t^2 z_N) - t^{Nd} g(z_1, \ldots, z_N)$ is a polynomial in $t$ with coefficients in $k[z_1, \ldots, z_N]$ and vanishes for all substitutions $t = b$ with $0 \neq b \in k$. Since $k$ is infinite, this polynomial is identically 0, i.e., $g(t^2 z_1, \ldots, t^2 z_N) = t^{Nd} g(z_1, \ldots, z_N)$. Next, choose a positive integer $m$ such that

$$f(t)^m g\left(\frac{tz_1}{z_1 + t}, \ldots, \frac{tz_N}{z_N + t}\right)$$

is in $k[z_1, \ldots, z_N][t]$. Then regard the polynomial

$$f(t)^{d+m} g\left(\frac{tz_1}{z_1 + t}, \ldots, \frac{tz_N}{z_N + t}\right) - t^{Nd} f(t)^m g(z_1, \ldots, z_N)$$

as a polynomial in the indeterminate $t$ with coefficients in $k[z_1, \ldots, z_N]$. From our assumption, it follows that the above polynomial vanishes for all substitutions $t = a$, where $0 \neq a \in k$. Since $k$ is infinite, this polynomial must be identically zero. Hence

$$g\left(\frac{tz_1}{z_1 + t}, \ldots, \frac{tz_N}{z_N + t}\right) = \frac{t^{Nd}}{f(t)^d} g(z_1, \ldots, z_N).$$

In this manner, we have shown that (ii) implies (iii).

Finally, assume (iii) holds. At the outset, note that by the assumed equation (1), $g$ is homogeneous of degree $Nd/2$ in $z_1, \ldots, z_N$. In particular, $Nd$ must be an even integer. Let $s, t, u, v$ be new indeterminates and $\lambda := \sqrt{sv - ut}$ (in a quadratic field-extension of $k(s, t, u, v)$). Letting $\alpha := \lambda/s$, the assumed properties (1) and (2) of $g$ imply

$$g\left(\alpha^2 z_1 + \frac{t}{s}, \ldots, \alpha^2 z_N + \frac{t}{s}\right) = \alpha^{Nd} g(z_1, \ldots, z_N).$$

Now let $\beta := s/u$ and for $1 \le j \le N$, substitute $\beta z_j/(z_j + \beta)$ in place of $z_j$ in the above equation. Then, using the assumed property (3) of $g$ and simplifying, we derive the equality

$$g\left(\frac{vz_1 + t}{uz_1 + s}, \ldots, \frac{vz_N + t}{uz_N + s}\right) = \frac{\lambda^{Nd}}{u^{Nd} f(s/u)^d} g(z_1, \ldots, z_N).$$

If $g = \phi(e_1, \ldots, e_N) \in k[e_1, \ldots, e_N]$, then as seen in the initial part of this proof, the left side of the above equality is simply $\phi(E_1, \ldots, E_N)$ and $A_0 = u^N f(s/u)$. So, defining $w := (Nd)/2$, we see that $w$ is a nonnegative integer and

$$A_0^d \phi(E_1, \ldots, E_N) = (sv - ut)^w \phi(e_1, \ldots, e_N).$$

Finally, observe that since each $E_j$ has degree 0 in $e_1, \ldots, e_N$ but $A_0$ has degree 1 in $e_1, \ldots, e_N$, the left side of the equation has degree at most $d$ in $e_1, \ldots, e_N$. Hence $\phi(e_1, \ldots, e_N)$ has degree at most $d$ in $e_1, \ldots, e_N$. Suppose $\phi(e_1, \ldots, e_N)$ has degree strictly less than $d$ in $e_1, \ldots, e_N$. Then the left side of the equation can be written in the form $A_0 \psi(A_0, \ldots, A_N)$, where $\psi(A_0, \ldots, A_N)$ is homogeneous as a polynomial in $A_0, \ldots, A_N$. Substituting $s = -z_1$ and $u = 1$ in the above equation, we note that $A_0 = f(-z_1) = 0$ and hence the left side of the above equation is zero. But $(-z_1 v - t)^w \ne 0$ and since $g \ne 0$, we have $\phi(e_1, \ldots, e_N) \ne 0$. This contradiction forces $\phi(e_1, \ldots, e_N)$ to be of degree exactly $d$ in $e_1, \ldots, e_N$. □

*Example* The most frequently encountered familiar binary invariant is the *discriminant*. If $F(X) := (X - z_1)(X - z_2) \cdots (X - z_N)$ (a monic polynomial in $X$ with roots $z_1, \ldots, z_N$), then

$$disc_X(F) := \prod_{1 \le i < j \le N} (z_i - z_j)^2$$

is, by definition, the discriminant (or $X$-discriminant) of $F$. It is straightforward to verify properties (1), (2) and (3) listed in (iii) of the above theorem; so, $disc_X(F)$ is indeed a binary invariant of type $(N, 2N - 2)$ (having weight $N(N - 1)$). Many more examples of this type follow.

**Definitions**  Let $k$ be a field.

1. Given an $N \times N$ matrix $A := [a_{ij}]$ with integer entries, let $r_i$ denote the sum of the entries in the $i$th row of $A$ for $1 \le i \le N$ and define

$$\rho(A) := (r_1, \ldots, r_N).$$

2. Given an $N \times N$ matrix $A := [a_{ij}]$, where each $a_{ij}$ is a nonnegative integer, letting $z$ stand for the vector $(z_1, \ldots, z_N)$, define

$$\delta(z, A) := \prod_{1 \le i < j \le N} (z_i - z_j)^{a_{ij}}.$$

3. Let $E(N)$ denote the set of all $N \times N$ symmetric matrices $A := [a_{ij}]$ such that each $a_{ij}$ is a nonnegative integer and $a_{ii} = 0$ for $1 \le i \le N$. For $V := (d_1, \ldots, d_N) \in \mathbb{Z}^N$, let $E(N, \le V)$ be the subset of $E(N)$ consisting of all $A \in E(N)$ such that letting $\rho(A) := (r_1, \ldots, r_N)$ we have $r_i \le d_i$ for $1 \le i \le N$. Let $E(N, V)$ be the subset consisting of $A \in E(N, \le V)$ such that $\rho(A) = V$. If $V = (d, d, \ldots, d)$, then $E(N, \le V)$ is denoted by $E(N, \le d)$ and $E(N, V)$ is denoted by $E(N, d)$.
4. Let $H_k(N, m)$ be the set of all $g \in k[z_1 - z_2, \ldots, z_1 - z_N]$ such that $g$ is homogeneous of degree $m$ in $z_1, \ldots, z_N$.
5. Let $V := (d_1, \ldots, d_N)$, where each $d_i$ is a nonnegative integer. Define $H_k(N, m, V)$ to be the set of all $g \in H_k(N, m)$ such that the $z_i$-degree of $g$ is $d_i$ for $1 \le i \le N$.

*Remarks*

1. If $V := (d_1, \ldots, d_N)$ and $E \in E(N, V)$, then observe that $\delta(z, E)$ is homogeneous of degree $m$ in $z_1, \ldots, z_N$, where $2m = d_1 + \cdots + d_N$ and the $z_i$-degree of $\delta(z, E)$ is $d_i$ for $1 \le i \le N$.
2. For $E(N, V)$ to be nonempty, it is necessary that $s := d_1 + \cdots + d_N$ is an even integer and $2d_i \le s$ for $1 \le i \le N$. Clearly, $E(2, V)$ is nonempty if and only if $d_1 = d_2$. In general, we are not aware of any conditions on $(N, V)$ that are known to be necessary and sufficient for $E(N, V)$ to be nonempty. Presently, we merely point out the following obvious fact. Suppose $d_1 + \cdots + d_N$ is an even integer. Let $P(N - 1, d_1)$ be the set of all ordered $(N - 1)$-tuples $a := (a_2, \ldots, a_N)$ of integers $a_i$ such that $0 \le a_i \le d_i$ for $2 \le i \le N$ and $a_2 + \cdots + a_N = d_1$. Given $a \in P(N - 1, d_1)$, let $W(a) := (d_2 - a_2, \ldots, d_N - a_N)$ and observe that the sum $(d_2 - a_2) + \cdots + (d_N - a_N)$ is also an even integer. Then the cardinality of $E(N, V)$ equals the sum of the cardinalities of $E(N - 1, W(a))$ as $a$ ranges over $P(N - 1, d_1)$.
3. A member of $E(N, d)$ can be thought of as the adjacency matrix of a $d$-regular (multi-) graph (without self-edges) on $N$ vertices. If $P$ is an $N \times N$ permutation matrix and $E \in E(N, d)$, then $PEP^{-1}$ is also in $E(N, d)$. Thus $S_N$ acts on $E(N, d)$ by conjugation. The resulting conjugacy classes are clearly in bijective correspondence with the isomorphism classes of $d$-regular (loopless, multi-) graphs on $N$

vertices. To the best of our knowledge, there is no known formula for counting the distinct (meaning pairwise nonisomorphic) $d$-regular (loopless, multi-) graphs on $N$ vertices (for general $d$ and $N$). For more on this enumeration and related recent results, the reader is referred to [1].

**Theorem 4** *Let $k$ be a field, let $N \geq 2$ be an integer and let $d_1, \ldots, d_N$ be nonnegative integers.*

(i) *Let $g \in H_k(N, m)$ be a nonzero polynomial such that the $z_i$-degree of $g$ is $d_i$ for $1 \leq i \leq N$. Then*

$$m \leq \frac{1}{2}(d_1 + \cdots + d_N).$$

(ii) *Let $V := (d_1, \ldots, d_N)$. Then, $H_k(N, m, V)$ is contained in the $k$-linear span of the set*

$$B(N, m, V) := \{\delta(z, E) \mid E \in E(N, \leq V)\} \cap H_k(N, m).$$

(iii) *Let $V := (d_1, \ldots, d_N)$. If $2m = d_1 + \cdots + d_N$, then $H_k(N, m, V)$ is contained in the $k$-linear span of the set*

$$\mu(N, V) := \{\delta(z, E) \mid E \in E(N, V)\}.$$

(iv) *Assume that $N!$ is nonzero in $k$. Then, $\mathcal{H}_k(m, d, N)$ is the $k$-linear span of the set*

$$\{Symm_N(\delta(z, E)) \mid E \in E(N, \leq d) \quad and \ \delta(z, E) \ has \ degree \ m\}.$$

*Proof* At the outset, we observe that $H_k(N, m)$ is exactly the set of polynomials that are homogeneous of degree $m$ as polynomials in $z_1 - z_2, \ldots, z_1 - z_N$. So, (i) follows from (iv) of Theorem 2.

We prove (ii) by lexicographic induction on $(N, d_N)$. If $N = 2$, then clearly $H_k(2, m, V)$ is empty unless $d_1 = d_2 = m$ and since

$$H_k(2, m, V) = \{c(z_1 - z_2)^m \mid c \in k\},$$

it is indeed a subset of the $k$-linear span of $B(2, m, V) = \{(z_1 - z_2)^m\}$. Henceforth, assume $N \geq 3$. Let $0 \neq g \in H_k(N, m, V)$. Let $e := d_N$ and

$$g = h_0 z_N^e + h_1 z_N^{e-1} + \cdots + h_e,$$

where $h_0 \neq 0$ and $h_i$ is homogeneous of degree $m - e + i$ in $z_1, \ldots, z_{N-1}$ for $0 \leq i \leq e$. Let $t$ be an indeterminate. Since $g(z_1 + t, \ldots, z_N + t) = g(z_1, \ldots, z_N)$, comparison of the coefficient of $z_N^e$ on both sides of this equation, we deduce that $h_0(z_1 + t, \ldots, z_{N-1} + t) = h_0(z_1, \ldots, z_{N-1})$. Hence

$$h_0 \in H_k(N - 1, m - e, W),$$

where $W := (d_1 - b_1, \ldots, d_{N-1} - b_{N-1})$ and $b_i$ is an integer such that $0 \le b_i \le d_i$ for $1 \le i \le N - 1$. By our induction hypothesis, $H_k(N - 1, m - e, W)$ is a subset of the $k$-linear span of $B(N - 1, m - e, W)$. In particular, if $d_N = 0$, then $b_i = 0$ for $1 \le i \le N - 1$ and we are done. Upon considering $E \in E(N - 1, W)$ such that $\delta(z, E)$ is homogeneous of degree $m - e$ in $z_1, \ldots, z_{N-1}$, one concludes

$$2(m - e) = 2(m - d_N) = (d_1 - b_1) + \cdots + (d_{N-1} - b_{N-1}).$$

Since $2m \le d_1 + \cdots + d_N$ by (i), we infer that $b_1 + \cdots + b_{N-1} \ge d_N$. Choose non-negative integers $a_j \le b_j$ for $1 \le j \le N - 1$ such that

$$a_1 + \cdots + a_{N-1} = e = d_N.$$

It follows that for any $\Delta \in B(N - 1, m - e, W)$, we clearly have

$$\delta := \Delta \prod_{j=1}^{N-1} (z_j - z_N)^{a_j} \in B(N, m, V)$$

and the $z_N$-degree of $\delta$ is exactly $d_N$. Consequently,

$$h := h_0 \prod_{j=1}^{N-1} (z_j - z_N)^{a_j}$$

is in the $k$-linear span of $B(N, m, V)$ and there is a $0 \ne c \in k$ such that the $z_N$-degree of $h_0 z_N^e - ch$ is at most $d_N - 1$. Now $g - ch \in H_k(N, m, U)$ for some $U := (q_1, \ldots, q_N)$ with $q_i \le d_i$ for $1 \le i \le N - 1$ and $q_N < d_N$. Then, by our induction hypothesis, $H_k(N, m, U)$ is contained in the $k$-linear span of $B(N, m, U)$. Since $B(N, m, U) \subseteq B(N, m, V)$, assertion (ii) is established.

To verify (iii), observe that if $E \in E(N, \le V) \setminus E(N, V)$, then by (i), the total degree of $\delta(z, E)$ in $z_1, \ldots, z_N$ is strictly less than $m$.

Assertion (iv) follows from the definition of $\mathcal{H}_k(m, d, N)$ and (iii).  □

*Remarks*

1. As noted earlier, there does not exist any nonzero semi-invariant of degree 1; hence $Symm_N(\delta(z, E)) = 0$ for all $0 \ne E \in E(N, \le 1)$.
2. For $k$ of characteristic 0, the above theorem is generally attributed to Cayley. Our treatment follows article no. 89 of [7].

**Definition**  Let $k$ be a field. Let $N$ and $d$ be positive integers with $N \geq 2$.

1. Let $\mathrm{Inv}_k(N, d)$ denote the $k$-linear subspace of $k[z_1, \ldots, z_N]$ consisting of all binary invariants of type $(N, d)$. Let $\mathrm{dinv}_k(N, d)$ denote the dimension of $\mathrm{Inv}_k(N, d)$ as a vector space over $k$.
2. Define
$$\mathfrak{G}(N, d, X) := \frac{(1 - X^{N+1})(1 - X^{N+2}) \cdots (1 - X^{N+d})}{(1 - X^2)(1 - X^3) \cdots (1 - X^d)}.$$

**Theorem 5** *Let $k$ be a field. Let $d$ and $N$ be positive integers such that $N \geq 2$.*

(i) *For $E \in E(N, d)$, the polynomial $Symm_N(\delta(z, E))$ is a binary invariant of type $(N, d)$.*

(ii) *Assume that $N!$ is a nonzero element of $k$. Then $Inv_k(N, d)$ is the $k$-linear span of*
$$M(N, d) := \{Symm_N(\delta(z, E)) \mid E \in E(N, d)\}.$$

(iii) *Assume that $N!$ is a nonzero element of $k$. Then,*
$$Inv_k(N, d) = \mathcal{H}_k(Nd/2, d, N).$$

(iv) *$\mathfrak{G}(N, d, X)$ and $\mathfrak{G}(N, d, X)/(1 - X)$ are polynomials in $X$ of degree $Nd + 1$ and $Nd$ respectively. For each integer $m$, $p(m, d, N)$ is the coefficient of $X^m$ in $\mathfrak{G}(N, d, X)/(1 - X)$. Moreover, we have*
$$\mathfrak{G}(d, N, X) = \mathfrak{G}(N, d, X) = (-1) \cdot X^{Nd+1} \cdot \mathfrak{G}(N, d, X^{-1}).$$

(v) *Assume that $\mathbb{Q} \subseteq k$ and $m$ is a nonnegative integer with $2m \leq Nd$. Then the dimension of $\mathcal{H}_k(m, d, N)$ (as a vector space over $k$) is the coefficient of $X^m$ in $\mathfrak{G}(N, d, X)$. In particular, $dinv_k(N, d)$ is the coefficient of $X^{Nd/2}$ in $\mathfrak{G}(N, d, X)$.*

(vi) *(Hermite Reciprocity) Assume that $\mathbb{Q} \subseteq k$. Then, $dinv_k(N, d) = dinv_k(d, N)$.*

*Proof*  If $Nd$ is odd, then both assertions hold trivially. Henceforth, assume $Nd$ to be an even integer. Fix an $E \in E(N, d)$, let $h(z_1, \ldots, z_N) := \delta(z, E)$ and $g := Symm_N(h)$. To prove (i), we proceed to verify (iii) of Theorem 3. Let $f(X)$ be as in Theorem 3. First note that $h$ is homogeneous of degree $w := Nd/2$ in $z_1, \ldots, z_N$ and hence
$$g(t^2 z_1, \ldots, t^2 z_N) = Symm_N(t^{Nd} h) = t^{Nd} g(z_1, \ldots, z_N).$$

Also, since $h$ is translation-invariant, we have
$$g = Symm_N(h(z_1 + t, \ldots, z_N + t)) = Symm_N(h(z_1, \ldots, z_N)).$$

Next, note that

$$\left(\frac{tz_i}{z_i + t} - \frac{tz_j}{z_j + t}\right)^{e_{ij}} = \left(\frac{t^2}{(z_i + t)(z_j + t)}\right)^{e_{ij}} (z_i - z_j)^{e_{ij}}$$

for all $1 \le i < j \le N$. Hence

$$h\left(\frac{tz_1}{z_1 + t}, \dots, \frac{tz_N}{z_N + t}\right) = \frac{t^{Nd}}{f(t)^d} h(z_1, \dots, z_N).$$

Since the first factor is symmetric in $z_1, \dots, z_N$, it follows that $g$ is a binary invariant of type $(N, d)$.

We proceed to prove (ii). By (i), each member of the $k$-linear span of $M(N, d)$ is a binary invariant of type $(N, d)$. Let $g \in k[z_1, \dots, z_N]$ be a nonzero binary invariant of type $(N, d)$. By (iii) of Theorem 2, the $z_i$-degree of $g$ is $d$ for $1 \le i \le N$. Also, from the remarks just preceding Theorem 3, $g$ is homogeneous of degree $w := Nd/2$. By (iii) of Theorem 3, $g \in k[z_1 - z_2, \dots, z_1 - z_N]$. Thus letting $V := (d, \dots, d)$, we have $g \in H_k(N, w, V)$. Now, by (iii) of Theorem 4, $g$ is in the $k$-linear span of the set $\mu(N, V)$. Recall that $E(N, V) = E(N, d)$ and hence $Symm_N(g)$ is in the $k$-linear span of the set $M(n, d)$. Since $g$ is symmetric in $z_1, \dots, z_N$, we have $Symm_N(g) = N! \, g$. Finally, $N!$ being a nonzero element of $k$, (ii) readily follows.

It suffices to prove (iii) with the additional assumption that $Nd$ is an even integer. The first remark following Theorem 2 together with (iv) of Theorem 4 imply that $\mathcal{H}_k(Nd/2, d, N)$ is the $k$-linear span of the set $M(N, d)$. So, by (ii), $\mathcal{H}_k(Nd/2, d, N) = \mathrm{Inv}_k(N, d)$.

To prove (iv), let $C_i(X)$ be the polynomials defined by the equation

$$\sum_{i \in \mathbb{N}} C_i(X) Y^i = \frac{1}{\prod_{j=0}^{N}(1 - X^j Y)} = \prod_{j=0}^{N} \left(\sum_{r=0}^{\infty} X^{jr} Y^r\right).$$

Observe that $p(m, d, N)$ is the coefficient of $X^m Y^d$ in the power-series expansion of the product on the extreme right. Clearly, each $C_i(X)$ is a polynomial in $X$ having integer coefficients (which are all positive). Also,

$$(1 - X^{N+1}Y) \cdot \sum_{i \in \mathbb{N}} C_i(X) X^i Y^i = \frac{(1 - X^{N+1}Y)}{\prod_{j=1}^{N+1}(1 - X^j Y)}$$

as well as

$$\frac{(1 - X^{N+1}Y)}{\prod_{j=1}^{N+1}(1 - X^j Y)} = (1 - Y) \cdot \sum_{i \in \mathbb{N}} C_i(X) Y^i.$$

From the above equations we infer that for $i \ge 1$,

$$(1 - X^i) \cdot C_i(X) = (1 - X^{N+i}) \cdot C_{i-1}(X)$$

and hence

$$\mathfrak{G}(N, i, X) = (1 - X) \cdot C_i(X).$$

Thus $\mathfrak{G}(N, d, X)/(1 - X)$ and $\mathfrak{G}(N, d, X)$ are polynomials in $X$ with integer coefficients; their degrees are easily seen to be $Nd$ and $Nd + 1$, respectively. Also, $p(m, d, N)$ is evidently the coefficient of $X^m$ in $\mathfrak{G}(N, d, X)/(1 - X)$. It is straightforward to verify that $\mathfrak{G}(N, d, X)$ satisfies the equations asserted in (iv).

Assuming $\mathbb{Q} \subseteq k$ and $2m \leq Nd$, assertion (vi) of Theorem 2 assures that as a vector space over $k$, $\mathcal{H}_k(m, d, N)$ has dimension $p(m, d, N) - p(m - 1, d, N)$. By (iv), $p(m, d, N)$ is the coefficient of $X^m$ in $\mathfrak{G}(N, d, X)/(1 - X)$ and hence $p(m, d, N) - p(m - 1, d, N)$ is the coefficient of $X^m$ in $\mathfrak{G}(N, d, X)$. Consequently, in view of (iii), $\mathrm{dinv}_k(N, d)$ is the coefficient of $X^{Nd/2}$ in $\mathfrak{G}(N, d, X)$. Thus (v) holds.

Lastly, (vi) follows from (v) and the equality $\mathfrak{G}(N, d, X) = \mathfrak{G}(d, N, X)$ established in (iv). $\qquad\qquad\square$

*Remarks*

1. From (ii) of Theorem 5, it follows that $\mathrm{Inv}_k(N, d)$ has a basis that is a subset of $\mathbb{Z}[z_1, \ldots, z_N]$.
2. We have $\mathfrak{G}(N, d, X) = \mathfrak{G}(N, d - 1, X) + X^d \mathfrak{G}(N - 1, d, X)$.
3. It is useful to observe that

$$\binom{N + d}{d}_q = \frac{\mathfrak{G}(N, d, q)}{(1 - q)},$$

where the left side stands for the (Gaussian) $q$-binomial coefficient.
4. For relatively small values of $N$ and $d$, the number $\mathrm{dinv}_k(N, d)$ may be large, for example: $\mathrm{dinv}_k(15, 18) = 8899$, $\mathrm{dinv}_k(20, 21) = 903256$.

**Theorem 6** *Let $N$, $d$ be positive integers both $\geq 2$. Let $k$ be a field containing $\mathbb{Q}$. Then $\mathrm{dinv}_k(N, d) = 0$ if and only if one of the following holds.*

   (i) *$Nd$ is an odd integer.*
  (ii) *$N = 2$ and $d$ is odd.*
 (iii) *$N$ is odd and $d = 2$.*
 (iv) *$N = 2 + 4m$ for some positive integer $m$ and $d = 3$.*
  (v) *$N = 3$ and $d = 2 + 4m$ for some positive integer $m$.*
 (vi) *$N = 5$ and $d \in \{6, 10, 14\}$.*
(vii) *$N = 6$ and $d \in \{3, 5, 7, 9, 11, 13\}$.*
(viii) *$N = 7$ and $d \in \{6, 10\}$.*
 (ix) *$N = 9$ and $d = 6$.*
  (x) *$N = 10$ and $d \in \{3, 5, 7\}$.*
 (xi) *$N = 11$ and $d = 6$.*
(xii) *$N = 13$ and $d = 6$.*
(xiii) *$N = 14$ and $d \in \{3, 5\}$.*

*Remarks*

1. The above Theorem 6 is established by J. Dixmier (see [8], as well as [9, Proposition 4.2]).
2. We mention two easily verified corollaries of Theorem 6. First is the fact that if $N \geq 15$, $d \geq 4$ (or if $N \geq 4$, $d \geq 15$) and $Nd$ is even, then $\mathrm{dinv}_k(N, d) \geq 1$. Second is the fact that if $N$ is an integer multiple of 4, then $\mathrm{dinv}_k(N, d) \geq 1$ for all $d \geq 2$ (or, reciprocally, if $d$ is an integer multiple of 4, then $\mathrm{dinv}_k(N, d) \geq 1$ for all $N \geq 2$).

## 2.3 Constructions of Semi-invariants

The polynomials and rational functions considered in this section have coefficients in an integral domain of characteristic zero. Since the notion of *u-adic order*, or simply *u-order* of a rational function plays a crucial role in our proofs, it is helpful to briefly recall a few basic definitions and properties related to *order*. As before, let $z_1, \ldots, z_N$ be indeterminates. Suppose $k$ is a unique factorization domain and $u$ is a prime element of the unique factorization domain $k[z_1, \ldots, z_N]$. Then, the *u-order* of a nonzero polynomial $h \in k[z_1, \ldots, z_N]$ is defined to be the largest nonnegative integer $m$ such that $h = u^m v$ for some $v \in k[z_1, \ldots, z_N]$; by convention, the *u*-order of 0 is $\infty$. More generally, for a rational function $f := P/Q$ with $P$, $Q$ in the polynomial ring $k[z_1, \ldots, z_N]$, the *u*-order of $f$ is defined to be the *u*-order of $P$ minus the *u*-order of $Q$. For any two rational functions $f$, $g$ in $z_1, \ldots, z_N$ with coefficients in $k$, it holds that the *u*-order of $fg$ is the sum of their respective *u*-orders whereas the *u*-order of $f + g$ is bounded below by the minimum of the *u*-orders of $f$ and $g$. Moreover, the *u*-order of $f + g$ equals the minimum of the *u*-orders of $f$ and $g$ whenever $f$ and $g$ have unequal *u*-orders.

**Definitions** Let $N \geq 2$ be an integer. As before, let $k$ be a field containing $\mathbb{Q}$, let $z_1, \ldots, z_N$ be indeterminates and let $z$ stand for $(z_1, \ldots, z_N)$.

1. For positive integers $m, n$, define $D_{(m,n)}$ to be the $m \times n$ matrix $[c_{ij}]$, where

$$c_{ii} := \begin{cases} 0 & \text{if } i = j, \\ 1 & \text{if } i \neq j. \end{cases}$$

   By $D_n$, we mean the $n \times n$ matrix $D_{(n,n)}$.
2. The *discriminant* $\Delta(z) \in \mathbb{Q}[z_1, \ldots, z_N]$ is defined by

$$\Delta(z) := \prod_{1 \leq i < j \leq N} (z_i - z_j)^2.$$

3. Let $m$ be a positive integer and let $\sigma \in S_m$ denote the $m$-cycle $(12 \cdots m)$. Given an ordered $m$-tuple

$$\mathfrak{a} := (a(1), \ldots, a(m)),$$

let cirmat($\mathfrak{a}$) denote the $m \times m$ *circulant* matrix $[c_{ij}]$ determined by $\mathfrak{a}$, i.e., for $1 \leq i, j \leq m$, let

$$c_{ij} := a(\sigma^{1-i}(j)).$$

4. Let $m, n$ be positive integers such that $mn = N$. Let $a, c$ be indeterminates. Let $\mathfrak{u} := (u(1), \ldots, u(m))$ be defined by

$$u(i) := \begin{cases} 2c & \text{if } 1 \leq i \leq \frac{m-1}{2}, \\ 0 & \text{otherwise.} \end{cases}$$

Let $M_0(m, n, a, c)$ be the $N \times N$ symmetric matrix defined as an $n \times n$ block-matrix $[M_{ij}]$, where, for $1 \leq i, j \leq n$,

$$M_{ij} := \begin{cases} 2aD_m & \text{if } i = j, \\ \text{cirmat}(\mathfrak{u}) & \text{if } i < j, \\ \text{cirmat}(\mathfrak{u})^T & \text{if } i > j. \end{cases}$$

*Examples*

$$M_0(3, 2, a, c) = \begin{bmatrix} 0 & 2a & 2a & 2c & 0 & 0 \\ 2a & 0 & 2a & 0 & 2c & 0 \\ 2a & 2a & 0 & 0 & 0 & 2c \\ 2c & 0 & 0 & 0 & 2a & 2a \\ 0 & 2c & 0 & 2a & 0 & 2a \\ 0 & 0 & 2c & 2a & 2a & 0 \end{bmatrix}$$

and

$$M_0(5, 2, a, c) = \begin{bmatrix} 2aD_5 & U \\ U^T & 2aD_5 \end{bmatrix},$$

where

$$U := \begin{bmatrix} 2c & 2c & 0 & 0 & 0 \\ 0 & 2c & 2c & 0 & 0 \\ 0 & 0 & 2c & 2c & 0 \\ 0 & 0 & 0 & 2c & 2c \\ 2c & 0 & 0 & 0 & 2c \end{bmatrix}.$$

Let $k$ be a field containing the rational numbers and let $z = (z_1, \ldots, z_N)$ be as in the above definitions. Recall that given an $N \times N$ matrix $A := [a_{ij}]$, the product $\prod(z_i - z_j)^{a_{ij}}$ ranging over $1 \leq i < j \leq N$ is denoted by $\delta(z, A)$. Below we review Theorem 2 of [10].

**Theorem 7**

(i) *Let $n$ be a positive integer and for $1 \leq i \leq n$, let $g_i \in \mathbb{Q}(z_1, \ldots, z_N)$ be such that $g_1 \neq 0$. Then, $g_1^2 + g_2^2 + \cdots + g_n^2 \neq 0$. In particular, given a $0 \neq g \in \mathbb{Q}(z_1, \ldots, z_N)$ and a nonempty subset $S \subseteq S_N$, we have*

$$\sum_{\sigma \in S} g(z_{\sigma(1)}, \ldots, z_{\sigma(N)})^2 \neq 0.$$

(ii) *Let $m, n, a, c$ be positive integers such that $3 \leq m \leq nm = N$ and $m$ is odd. Then, letting $M_0 := M_0(m, n, a, c)$, we have $M_0 \in E(N, (2a + cn - c)(m - 1))$ and $Symm_N(\delta(z, M_0)) \neq 0$.*

*Proof* To prove (i), let $h := g_1^2 + g_2^2 + \cdots + g_n^2$. For $1 \leq i \leq n$, let $p_i, q_i \in \mathbb{Q}[z_1, \ldots, z_N]$ be polynomials such that $g_i q_i = p_i$ and $q_i \neq 0$. Note that $g_1 \neq 0$ implies $p_1 \neq 0$. Now since $f := p_1 q_1 q_2 \cdots q_n$ is a nonzero polynomial, there exists $(a_1, \ldots, a_N) \in \mathbb{Q}^N$ such that $f(a_1, \ldots, a_N) \neq 0$. Fix such $(a_1, \ldots, a_N)$ and let $c_i := g_i(a_1, \ldots, a_N)$ for $1 \leq i \leq n$. Then each $c_i$ is a rational number and $c_1 \neq 0$. Since $c_1^2 > 0$ and $(c_2^2 + \cdots + c_n^2) \geq 0$, we have $h(a_1, \ldots, a_N) > 0$. In particular, $h \neq 0$. This proves (i).

To prove (ii), let $M_{ij}$ denote the $ij$th $m \times m$ block of $M_0$ (as in the definition of $M_0$). If $1 \leq i < j \leq n$, then $M_{ij}$ being a circulant matrix and $m$ being odd, each row-sum as well as each column-sum of $M_{ij}$ is exactly $c(m - 1)$. Now it is easily verified that $M_0$ is a member of $E(N, (2a + cn - c)(m - 1))$. Since each entry of $M_0$ is a nonnegative even integer, there exists a nonzero polynomial $g \in k[z_1, \ldots, z_N]$ such that

$$Symm_N(\delta(z, M_0)) = \sum_{\sigma \in S_N} \sigma(g(z_1, \ldots, z_N))^2.$$

Therefore, (ii) follows from (i).                                                           □

**Definitions** Let $N$ be an integer such that $2 \leq N$. As before, let $k$ be a field containing $\mathbb{Q}$, let $z_1, \ldots, z_N$ be indeterminates and let $z$ stand for $(z_1, \ldots, z_N)$.

1. Given a subset $B$ of $\{1, 2, \ldots, N\}$, let

$$\pi(B) := \{(i, j) \in B \times B \mid i < j\}.$$

The set $\pi(B)$ is tacitly identified with the set of all 2-element subsets of the set $B$, i.e.,

$$\pi(B) = \{\{i, j\} \mid i, j \in B \text{ and } i \neq j\}.$$

By $\pi[N]$, we mean the set $\pi(\{1, \ldots, N\})$.

2. Given a subset $C \subseteq \pi[N]$ and a function $\varepsilon : C \to \mathbb{N}$, the image of $(i, j) \in C$ via $\varepsilon$ is denoted by $\varepsilon(i, j)$. A nonnegative integer $w$ is identified with the constant function $C \to \mathbb{N}$ that maps each member of $C$ to $w$.

3. For a subset $C \subseteq \pi[N]$ and a function $\varepsilon : C \to \mathbb{N}$, define

$$v(z, C, \varepsilon) := \prod_{(i,j)\in C} (z_i - z_j)^{\varepsilon(i,j)}.$$

By convention, $v(z, \emptyset, \varepsilon) = 1$.

4. Let $p$ be a positive integer and $\mathfrak{n} : 0 = n_0 < n_1 < \cdots < n_p = N$ be a sequence of integers. For $1 \le r \le p$, let

$$S_r(\mathfrak{n}) := \{i \mid n_{r-1} + 1 \le i \le n_r\}$$

and for $i \in S_r(\mathfrak{n})$, define

$$\chi(\mathfrak{n}, i) := |\{s \mid 1 \le s \le p, \ (n_s - n_{s-1}) < (i - n_{r-1})\}|,$$
$$w_i(\mathfrak{n}) := (2a - c)(n_r - n_{r-1} - 1) + c(N + \chi(\mathfrak{n}, i) - p).$$

Let $a, c$ be indeterminates and let $M(\mathfrak{n}, a, c)$ denote the $N \times N$ symmetric matrix $[u(i,j)]$ whose upper-triangular entries are defined by

$$u(i,j) := \begin{cases} 2a & \text{if } (i,j) \in \pi(S_r(\mathfrak{n})), \\ 0 & \text{if } (i,j) = (\epsilon + n_{r-1}, \ \epsilon + n_{s-1}) \text{ with } r < s, \\ c & \text{otherwise.} \end{cases}$$

5. Let $m, n$ be positive integers such that $m \ge 2$. Let $a, c$ be indeterminates. Let $M(m, n, a, c)$ denote the $(mn) \times (mn)$ symmetric matrix $[a(i,j)]$ whose entries are defined by

$$a(i,j) := \begin{cases} 2a & \text{if } l_1 = l_2 \text{ and } r_1 \ne r_2, \\ 0 & \text{if } r_1 = r_2, \\ c & \text{otherwise,} \end{cases}$$

where $l_1, l_2, r_1, r_2$ are integers such that $(i,j) = (l_1 m + r_1, \ l_2 m + r_2)$, $0 \le l_1, l_2 \le n - 1$ and $1 \le r_1, r_2 \le m$.

*Remarks*

1. There is an obvious bijective correspondence between functions $\varepsilon$ from $\pi[N]$ to $\mathbb{N}$ and $N \times N$ symmetric matrices $[a_{ij}]$ having

$$a_{ii} = 0 \quad \text{for } 1 \le i \le N$$

given by the prescription

$$a_{ij} = a_{ji} = \varepsilon(i, j) \quad \text{for } 1 \le i < j \le N.$$

2. Given an integer sequence

$$\mathfrak{n} : 0 = n_0 < n_1 < \cdots < n_p = N,$$

and integers $a$, $c$ as in the above definition, $M(\mathfrak{n}, a, c)$ is realized as a $p \times p$ block-matrix $[M_{ij}]$, where

$$M_{rr} = 2aD_{n_r - n_{r-1}} \quad \text{for } 1 \le r \le p$$

and if $1 \le r \ne s \le p$, then

$$M_{sr}^T = M_{rs} := cD_{(n_r - n_{r-1}, \, n_s - n_{s-1})}.$$

3. Likewise, $M(m, n, a, c)$ is seen to be an $n \times n$ block-matrix $[M_{ij}]$, where

$$M_{rr} = 2aD_m \quad \text{for } 1 \le r \le n,$$

and if $1 \le r \ne s \le p$, then $M_{sr}^T = M_{rs} := cD_m$. Some examples follow.

*Examples*

1. Let $N = 4$, $p = 3$ and $\mathfrak{n} := 0 < 1 < 3 < 4$. Then $M(\mathfrak{n}, a, c)$ can be viewed in a block-format, where the block-sizes are $1 = 1 - 0$, $2 = 3 - 1$ and $1 = 4 - 3$:

$$M(\mathfrak{n}, a, c) = \begin{bmatrix} 0 & 0 & c & 0 \\ 0 & 0 & 2a & 0 \\ c & 2a & 0 & c \\ 0 & 0 & c & 0 \end{bmatrix}.$$

Suppose $2a$ and $c$ are positive integers and view $M(\mathfrak{n}, a, c)$ as the adjacency-matrix (or the edge-matrix) of a undirected, loopless multi-graph on 4 vertices labeled by $1, 2, 3, 4$. Then, from vertex 1 there are exactly $c$ edges all of which terminate in vertex 3, from vertex 2 there are exactly $2a$ edges all of which terminate in vertex 3, etc. By changing vertex-label 1 to 2, vertex-label 2 to 3 and vertex-label 3 to 1, we obtain an isomorphic multi-graph whose adjacency matrix is in the middle on the right side of the following equation:

$$M(\mathfrak{n}, a, c) = \begin{bmatrix} 0 & 1 & 0 & 0 \\ 0 & 0 & 1 & 0 \\ 1 & 0 & 0 & 0 \\ 0 & 0 & 0 & 1 \end{bmatrix} \begin{bmatrix} 0 & c & 2a & c \\ c & 0 & 0 & 0 \\ 2a & 0 & 0 & 0 \\ c & 0 & 0 & 0 \end{bmatrix} \begin{bmatrix} 0 & 0 & 1 & 0 \\ 1 & 0 & 0 & 0 \\ 0 & 1 & 0 & 0 \\ 0 & 0 & 0 & 1 \end{bmatrix}.$$

2. For $N = 8$ and $m = 4$, we have

$$
M(4, 2, a, c) := \left[\begin{array}{cccc|cccc}
0 & 2a & 2a & 2a & 0 & c & c & c \\
2a & 0 & 2a & 2a & c & 0 & c & c \\
2a & 2a & 0 & 2a & c & c & 0 & c \\
2a & 2a & 2a & 0 & c & c & c & 0 \\
\hline
0 & c & c & c & 0 & 2a & 2a & 2a \\
c & 0 & c & c & 2a & 0 & 2a & 2a \\
c & c & 0 & c & 2a & 2a & 0 & 2a \\
c & c & c & 0 & 2a & 2a & 2a & 0
\end{array}\right].
$$

Consider the multi-graph on vertices $1, \dots, 8$ having adjacency matrix $M(4, 2, a, c)$. Label vertices $2, 3, 4, 5, 6, 7$ as $3, 5, 7, 2, 4, 6$ respectively to obtain an isomorphic multi-graph. Then the adjacency matrix of this later multi-graph is:

$$
\left[\begin{array}{cc|cc|cc|cc}
0 & 0 & 2a & c & 2a & c & 2a & c \\
0 & 0 & c & 2a & c & 2a & c & 2a \\
\hline
2a & c & 0 & 0 & 2a & c & 2a & c \\
c & 2a & 0 & 0 & c & 2a & c & 2a \\
\hline
2a & c & 2a & c & 0 & 0 & 2a & c \\
c & 2a & c & 2a & 0 & 0 & c & 2a \\
\hline
2a & c & 2a & c & 2a & c & 0 & 0 \\
c & 2a & c & 2a & c & 2a & 0 & 0
\end{array}\right].
$$

It is straightforward to verify that the above matrix is the product

$$
P \cdot M(4, 2, a, c) \cdot P^{T},
$$

where $P$ is the $8 \times 8$ permutation-matrix corresponding to the product of disjoint cycles: $(2\,3\,5)(4\,7\,6)$.

Now we focus our attention on a special type of $N \times N$ symmetric matrix with nonnegative integer entries. Firstly, each of these is expressible as a $q \times q$ ($q$, a positive integer) block-matrix $[C_{ij}]$, where the size of $C_{rr}$ is $m_r \times m_r$ and the block-sizes are nondecreasing positive integers, i.e., $1 \le m_r \le m_{r+1}$ for $1 \le r < q$. For convenience, let $m_0 = 0$. Then, clearly

$$
0 = m_0 < m_1 \le m_2 \le \cdots \le m_q \quad \text{and} \quad \sum_{i=1}^{q} m_i = N.
$$

Secondly, we require each diagonal block to be a zero matrix, i.e., $C_{rr} = 0$ for $1 \le r \le q$. Since we consider only the symmetric matrices, $C_{rs} = C_{sr}^{T}$ for $1 \le r \le s \le q$. Such a matrix is completely determined by the entries in its strict upper-triangle.

So, as seen in the above definitions, it suffices to consider the function $\varepsilon : \pi[N] \to \mathbb{N}$ corresponding to the matrix. Corresponding to the sequence $\mathfrak{m} : 0 = m_0 < m_1 \leq m_2 \leq \cdots \leq m_q$, define

$$A_r(\mathfrak{m}) := \left\{ i + \sum_{j=0}^{r-1} m_j \;\middle|\; 1 \leq i \leq m_r \right\}$$

for $1 \leq r \leq q$. Then, for $1 \leq r < s \leq q$, the block $C_{rs}$ (which is a $m_r \times m_s$ matrix) consists of entries $\varepsilon(i, j)$ with $(i, j) \in A_r(\mathfrak{m}) \times A_s(\mathfrak{m})$. Also, note that $\varepsilon(i, j) = 0$ for all $(i, j) \in \pi(A_r(\mathfrak{m}))$, $1 \leq r \leq q$. Now we define a special property which plays a crucial role in ensuring that $Symm_N(v(z, \pi[N], \varepsilon))$ is a nonzero polynomial.

**Definition** Let $\mathfrak{m}$ be as above and let $\varepsilon : \pi[N] \to \mathbb{N}$ be a function. We say $\varepsilon$ is $\mathfrak{m}$-*excellent* provided it satisfies the property (1) below and at least one of the two properties (2) and (3) that follow.

(1) $\varepsilon(i, j) = 0$ if and only if $(i, j) \in \pi(A_r(\mathfrak{m}))$ with $1 \leq r \leq q$.
(2) For $1 \leq r < s \leq q$, there is a nonnegative integer $b(m_r, m_s)$, depending only on $(m_r, m_s)$, such that

$$\sum_{(i,j) \in A_r(\mathfrak{m}) \times A_s(\mathfrak{m})} \varepsilon(i, j) = b(m_r, m_s), \quad \text{and}$$

if $m_r = m_s$, then $b(m_r, m_s)$ is an even integer.

(3) For $1 \leq r < s \leq q$,

$$\sum_{(i,j) \in A_r(\mathfrak{m}) \times A_s(\mathfrak{m})} \varepsilon(i, j) \quad \text{is an even integer.}$$

*Remarks* Fix $\mathfrak{m}$ as above and let *excellence* mean $\mathfrak{m}$-excellence.

1. Property (1) in the definition of excellence is equivalent to the requirement that $C_{rr} = 0$ for $1 \leq r \leq q$ and each entry of $C_{rs}$ is positive if $1 \leq r < s \leq q$.
2. Given a matrix $B$, let $\|B\|$ denote the sum of all entries of $B$. Then, property (2) in the definition of excellence can be paraphrased as follows: if matrices $C_{ij}$ and $C_{rs}$ are of the same size (i.e., if $m_i = m_r$ and $m_j = m_s$), then $\|C_{ij}\| = \|C_{rs}\|$ and if $C_{rs}$ is a square matrix (i.e., if $m_r = m_s$), then $\|C_{rs}\|$ is an even integer.
3. Property (3) in the definition of excellence is satisfied if and only if $\|C_{rs}\|$ is an even integer for all $1 \leq r \leq s \leq q$.
4. Note that if $m_1 < m_2 < \cdots < m_q$, then the property (2) in the definition of excellence always holds (it is vacuously satisfied).
5. It is important to observe that the properties (2) and (3) in the definition of excellence are independent, i.e., while any one of them is satisfied the other may fail to hold; the following examples illustrate this fact.

*Examples*  Consider the symmetric matrices $M_1$ and $M_2$ below.

$$M_1 := \begin{bmatrix} 0 & 2 & 1 & 4 \\ 2 & 0 & 2 & 3 \\ 1 & 2 & 0 & 0 \\ 4 & 3 & 0 & 0 \end{bmatrix}, \qquad M_2 := \begin{bmatrix} 0 & 2 & 1 & 3 \\ 2 & 0 & 2 & 4 \\ 1 & 2 & 0 & 0 \\ 3 & 4 & 0 & 0 \end{bmatrix}.$$

Note that we have $q = 3$, $m_1 = 1 = m_2$ and $m_3 = 2$. Let $\varepsilon_i : \pi[4] \to \mathbb{N}$ be the function associated with $M_i$ for $i = 1, 2$. Then, $\varepsilon_1$ satisfies properties (1) and (2) but does not satisfy property (3). On the other hand, $\varepsilon_2$ satisfies properties (1) and (3) but does not satisfy property (2).

**Definitions**  We continue to use the above notation. In particular, let m be the integer sequence $0 = m_0 < m_1 \le m_2 \le \cdots \le m_q$ with $m_1 + \cdots + m_q = N$.

1. Define
$$\pi[\mathrm{m}] := \bigcup_{1 \le r < s \le q} A_r(\mathrm{m}) \times A_s(\mathrm{m}).$$

2. For $\theta \in S_N$ and $(i, j) \in \pi[N]$, let
$$\theta(i, j) := \begin{cases} (\theta(i),\ \theta(j)) & \text{if } \theta(i) < \theta(j), \\ (\theta(j),\ \theta(i)) & \text{if } \theta(j) < \theta(i). \end{cases}$$

3. For $\theta \in S_N$ and $1 \le r \le q$, let
$$B_r(\mathrm{m},\ \theta) := \{i \mid 1 \le i \le N \ \text{and} \ \theta(i) \in A_r(\mathrm{m})\}.$$

4. For $\theta \in S_N$, let
$$R(\mathrm{m},\ \theta) := \bigcup_{1 \le r \le q} \pi\,(B_r(\mathrm{m},\ \theta)).$$

5. Let $G(\mathrm{m}) := \{\theta \in S_N \mid \theta(i, j) \in \pi[\mathrm{m}] \ \text{for all} \ (i, j) \in \pi[\mathrm{m}]\}$.

*Remarks*

1. Note that
$$\pi[\mathrm{m}] = \pi[N] \setminus \bigcup_{i=1}^{q} \pi(A_i(\mathrm{m})).$$

2. The second of the above definitions prescribes an action of $S_N$ on $\pi[N]$; henceforth, we tacitly identify $S_N$ as a subgroup of the group of permutations of $\pi[N]$ via this action.

3. Since $B_i(\mathrm{m},\ \theta) = \theta^{-1}(A_i(\mathrm{m}))$, it is evident that $B_i(\mathrm{m},\ \theta)$ has cardinality $m_i$ for $1 \le i \le q$ and the sets $B_1(\mathrm{m},\ \theta), \ldots, B_q(\mathrm{m},\ \theta)$ partition $\{1, \ldots, N\}$.

*Example* Let $N = 4$, $q = 3$, and $\mathfrak{m} : 0 < 1 \leq 1 < 3$. Then, $A_1(\mathfrak{m}) = \{1\}$, $A_2(\mathfrak{m}) = \{2\}$, $A_3(\mathfrak{m}) = \{3, 4\}$ and $\pi[\mathfrak{m}] = \{(1, 2), (1, 3), (1, 4), (2, 3), (2, 4)\}$. Observe that $G(\mathfrak{m})$ consists of 4 permutations: namely, the identity permutation, the transpositions $(1, 2)$, $(3, 4)$ and their product $\theta := (1, 2)(3, 4)$. Consequently, we have $B_1(\mathfrak{m}, \theta) = \{2\} = A_2(\mathfrak{m})$, $B_2(\mathfrak{m}, \theta) = \{1\} = A_1(\mathfrak{m})$ and $B_3(\mathfrak{m}, \theta) = \{3, 4\} = A_3(\mathfrak{m})$. Also, $R(\mathfrak{m}, \theta) = \{(3, 4)\}$. Now $\eta \in S_4 \setminus G(\mathfrak{m})$ if and only if $\{\eta(3), \eta(4)\} \neq \{3, 4\}$ if and only if $R(\mathfrak{m}, \eta) \neq \{(3, 4)\}$ if and only if $\pi \cap R(\mathfrak{m}, \eta) \neq \emptyset$.

**Lemma 2** *We continue to use the above notation. Fix a sequence* $\mathfrak{m}$ *as above with* $q \geq 2$ *and fix a* $\theta \in G(\mathfrak{m})$. *Then, given* $r$ *with* $1 \leq r \leq q$, *there is a unique integer* $r(\mathfrak{m}, \theta)$ *such that the following holds.*

(i) *We have* $1 \leq r(\mathfrak{m}, \theta) \leq q$ *and* $B_r(\mathfrak{m}, \theta) = A_{r(\mathfrak{m},\theta)}(\mathfrak{m})$. *In particular, the sequence* $\theta(A_1(\mathfrak{m})), \ldots, \theta(A_q(\mathfrak{m}))$ *is a permutation of* $A_1(\mathfrak{m}), \ldots, A_q(\mathfrak{m})$.

(ii) *For* $1 \leq r < s \leq q$, *we have*

$$\pi[\mathfrak{m}] \cap \left(A_{r(\mathfrak{m},\theta)} \times A_{s(\mathfrak{m},\theta)}\right) \neq \emptyset \quad \text{if and only if } r(\mathfrak{m}, \theta) < s(\mathfrak{m}, \theta).$$

(iii) *For* $1 \leq r \leq q$, *we have* $m_{r(\mathfrak{m},\theta)} = m_r$.

(iv) *Let* $\theta \in S_N$. *Then,* $\pi[\mathfrak{m}] \cap R(\mathfrak{m}, \theta) = \emptyset$ *if and only if* $\theta \in G(\mathfrak{m})$.

*Proof* Since $\theta \in G(\mathfrak{m})$, given an ordered pair $(i, j) \in \pi(A_r(\mathfrak{m}))$ with $1 \leq r \leq q$, it is clear that there is a unique $s$ with $1 \leq s \leq q$ such that $\theta(i, j) \in \pi(A_s(\mathfrak{m}))$. For an $s$ with $1 \leq s \leq q$, choose $i \in B_r(\mathfrak{m}, \theta) \cap A_s(\mathfrak{m})$. Then, for any $j \in A_s(\mathfrak{m})$ with $j \neq i$, we must have $\theta(i, j) \in \pi(A_r(\mathfrak{m}))$ and hence $j \in B_r(\mathfrak{m}, \theta)$. It follows that $A_s(\mathfrak{m}) \subseteq B_r(\mathfrak{m}, \theta)$. If $1 \leq s < p \leq q$ are such that $A_s(\mathfrak{m}) \cup A_p(\mathfrak{m}) \subseteq B_r(\mathfrak{m}, \theta)$, then an $(i, j) \in A_s(\mathfrak{m}) \times A_p(\mathfrak{m})$ is in $\pi$ whereas $\theta(i, j)$ is in $\pi(A_r(\mathfrak{m}))$. But this is impossible since $\theta \in G(\mathfrak{m})$. Thus we have established assertion (i). Observe that $r(\mathfrak{m}, \theta) \neq s(\mathfrak{m}, \theta)$ for $1 \leq r < s \leq q$ and hence assertion (ii) holds. Since $B_r(\mathfrak{m}, \theta)$ has cardinality $m_r$, the equality $B_r(\mathfrak{m}, \theta) = A_{r(\mathfrak{m},\theta)}$ established in (i) verifies assertion (iii). To prove (iv), observe that $\pi[N]$ is partitioned by the sets $\theta^{-1}(\pi[\mathfrak{m}])$ and $R(\mathfrak{m}, \theta)$. So, $\pi[\mathfrak{m}] \cap R(\mathfrak{m}, \theta) = \emptyset$ if and only if $\pi[\mathfrak{m}] = \theta^{-1}(\pi[\mathfrak{m}])$ if and only if $\theta \in G(\mathfrak{m})$. □

Before proceeding to the general setting, it is helpful to consider a concrete example which illustrates the role played by the excellence of $\varepsilon$ in ensuring nontriviality of $Symm_N(v(z, \pi[N], \varepsilon))$. Let us consider the case of $N = 4$, $q = 3$, $m_1 = 1 = m_2$ and $m_3 = 2$. Assume $\varepsilon(i, j) = 0$ if and only if $(i, j) = (3, 4)$. So, $\mu := v(z, \pi[4], \varepsilon)$ is given by

$$\mu = (z_1 - z_2)^{\varepsilon(1,2)}(z_1 - z_3)^{\varepsilon(1,3)}(z_1 - z_4)^{\varepsilon(1,4)}(z_2 - z_3)^{\varepsilon(2,3)}(z_2 - z_4)^{\varepsilon(2,4)}.$$

To explore whether $Symm_4(\mu)$ is nonzero, firstly, we substitute $z_1 = tx_1 + t_1$, $z_2 = tx_2 + t_2$, $z_3 = tx_3 + t_3$, $z_4 = tx_4 + t_3$ in $Symm_4(\mu)$, where $t, x_1, x_2, x_3, x_4$ and $t_1, t_2, t_3$ are new indeterminates. After this substitution, we examine each summand in $Symm_4(\mu)$ for divisibility by $t$. Recall, from the example just prior to Lemma 2, that in our case the associated $G(\mathrm{m})$ has only four members: identity, transposition $(1, 2)$, transposition $(3, 4)$ and $\theta := (1, 2)(3, 4)$. Applying each of these to $\mu$ respectively, we obtain $\mu_1 := \mu$,

$$\mu_2 := (z_2 - z_1)^{\varepsilon(1,2)}(z_2 - z_3)^{\varepsilon(1,3)}(z_2 - z_4)^{\varepsilon(1,4)}(z_1 - z_3)^{\varepsilon(2,3)}(z_1 - z_4)^{\varepsilon(2,4)},$$
$$\mu_3 := (z_1 - z_2)^{\varepsilon(1,2)}(z_1 - z_4)^{\varepsilon(1,3)}(z_1 - z_3)^{\varepsilon(1,4)}(z_2 - z_4)^{\varepsilon(2,3)}(z_2 - z_3)^{\varepsilon(2,4)},$$
$$\mu_4 := (z_2 - z_1)^{\varepsilon(1,2)}(z_2 - z_4)^{\varepsilon(1,3)}(z_2 - z_3)^{\varepsilon(1,4)}(z_1 - z_4)^{\varepsilon(2,3)}(z_1 - z_3)^{\varepsilon(2,4)}.$$

Since $\mu_i$ is not a multiple of $(z_3 - z_4)$, after the aforementioned substitution, $\mu_i$ is not divisible by $t$ for $1 \leq i \leq 4$. Moreover, letting $t = 0$ after the substitution, $\mu_1$, $\mu_2$, $\mu_3$, $\mu_4$ are transformed respectively to $\nu_1, \nu_2, \nu_3, \nu_4$, where $\nu_3 = \nu_1, \nu_4 = \nu_2$,

$$\nu_1 := (t_1 - t_2)^{\varepsilon(1,2)}(t_1 - t_3)^{\varepsilon(1,3)+\varepsilon(1,4)}(t_2 - t_3)^{\varepsilon(2,3)+\varepsilon(2,4)},$$
$$\nu_2 := (t_2 - t_1)^{\varepsilon(1,2)}(t_2 - t_3)^{\varepsilon(1,3)+\varepsilon(1,4)}(t_1 - t_3)^{\varepsilon(2,3)+\varepsilon(2,4)}.$$

Observe that $\varepsilon$ satisfies the excellence property (2) if and only if $\varepsilon(1, 2)$ is even and $\varepsilon(1, 3) + \varepsilon(1, 4) = \varepsilon(2, 3) + \varepsilon(2, 4)$ and in such a case we have $\nu_1 + \nu_2 + \nu_3 + \nu_4 = 4\nu_1$. On the other hand, $\varepsilon$ satisfies the excellence property (3) if and only if each of $\varepsilon(1, 2)$, $\varepsilon(1, 3) + \varepsilon(1, 4)$, $\varepsilon(2, 3) + \varepsilon(2, 4)$ is even. Thus excellence of $\varepsilon$ ensures that $\nu_1 + \nu_2 + \nu_3 + \nu_4 \neq 0$. Next, consider a permutation $\eta \in S_4 \setminus G(\mathrm{m})$. Then, there is some $(i, j) \in \pi[4]$ such that $(i, j) \neq (3, 4)$ and $\eta(i, j) = (3, 4)$. For this $(i, j)$, excellence property (1) satisfied by $\varepsilon$ makes sure that $\varepsilon(i, j) \neq 0$ and so, $\eta(\mu)$ is divisible by $t(x_3 - x_4)$. It now follows that after our substitution, $Symm_4(\mu) - 4\nu_1$ is divisible by $t$, assuming property (2). Of course, since $\nu_1$ is nonzero and independent of the variable $t$, our $Symm_4(\mu)$ must be nonzero to begin with. The same conclusion is reached if property (3) holds. For treating the case of general $N$ and $\mathrm{m}$, we need some preparatory definitions and a lemma.

**Definitions** Let $\mathrm{m} : 0 < m_1 \leq \cdots \leq m_q$ be as above. Assume $q \geq 2$. Let $t, t_1, \ldots,$ $t_q, x_1, \ldots, x_N$ be indeterminates. Let $x$ stand for $(x_1, \ldots, x_N)$, $T$ stand for $(t_1, \ldots, t_q)$ and let $k[t, T, x]$ stand for $k[t, t_1, \ldots, t_q, x_1, \ldots, x_N]$.

1. Given $f \in k[t, T, X]$, by the $x$-*degree* (resp. $T$-*degree*) of $f$, we mean the total degree of $f$ in the indeterminates $x_1, \ldots, x_N$ (resp. $t_1, \ldots, t_q$).
2. Define $\phi : k[z] \to k[t, T, x]$ to be the injective $k$-homomorphism of rings given by

$$\phi(z_i) := tx_i + t_r, \quad \text{if } i \in A_r(\mathrm{m}) \text{ with } 1 \leq r \leq q.$$

3. For $\theta \in G(\mathrm{m})$, define $V_\theta(\mathrm{m}, \varepsilon) := \phi(\theta(v(z, \pi[\mathrm{m}], \varepsilon)))$ and let

$$V(\mathrm{m}, \varepsilon) := \sum_{\theta \in G(\mathrm{m})} V_\theta(\mathrm{m}, \varepsilon).$$

4. Let $d(\mathrm{m}) := \sum_{(i,j) \in \pi[\mathrm{m}]} \varepsilon(i, j)$.

**Lemma 3** *Fix* m $: 0 < m_1 \leq \cdots \leq m_q$ *as above. Assume* $q \geq 2$.

(i) *Given* $\theta \in S_N$, $(i, j) \in \pi[N]$ *and* $1 \leq r, s \leq q$, *we have*

$$\phi(z_{\theta(i)} - z_{\theta(j)}) = t(x_{\theta(i)} - x_{\theta(j)}) + (t_r - t_s)$$

*if and only if* $(\theta(i), \theta(j)) \in A_r(\mathrm{m}) \times A_s(\mathrm{m})$. *In particular,*

$$\phi(z_{\theta(i)} - z_{\theta(j)}) = t(x_{\theta(i)} - x_{\theta(j)}) \quad \textit{if and only if} \quad (i, j) \in R(\mathrm{m}, \theta).$$

(ii) *Suppose* $\varepsilon$ *is* m-*excellent and* $\theta \in G(\mathrm{m})$. *Let* $f_\theta(t, T, x) := V_\theta(\mathrm{m}, \varepsilon)$. *Then,* $f_\theta(0, T, x) \neq 0$, $f_\theta(0, T, x)$ *is homogeneous in* $T$ *and the* $T$-*degree of* $f_\theta(0, T, x)$ *is* $d(\mathrm{m})$.

(iii) *Suppose* $\varepsilon$ *is* m-*excellent. Let* $f(t, T, x) := V(\mathrm{m}, \varepsilon)$. *Then,* $f(0, T, x) \neq 0$, $f(0, T, x) \neq 0$ *is homogeneous in* $T$ *of* $T$-*degree* $d(\mathrm{m})$.

*Proof* Assertion (i) can be verified in a straightforward manner. We proceed to prove (ii). For an ordered pair $(r, s)$ with $1 \leq r, s \leq q$, define

$$S_\theta(\mathrm{m}, r, s) := \pi[\mathrm{m}] \cap (A_{r(\mathrm{m}, \theta)} \times A_{s(\mathrm{m}, \theta)})$$

and note that $S_\theta(\mathrm{m}, r, s) = \emptyset$ if $s(\mathrm{m}, \theta) < r(\mathrm{m}, \theta)$. It is straightforward to verify that

$$f_\theta(0, T, x) = \prod_{1 \leq r < s \leq q} \left( \prod_{(i,j) \in S_\theta(\mathrm{m},r,s)} (t_r - t_s)^{\varepsilon(i,j)} \cdot \prod_{(i,j) \in S_\theta(\mathrm{m},s,r)} (t_s - t_r)^{\varepsilon(i,j)} \right).$$

Firstly, suppose the m-excellence property (2) holds. Then, for $1 \leq r < s \leq q$, we have

$$\sum_{(i,j) \in S_\theta(\mathrm{m},r,s)} \varepsilon(i, j) = \begin{cases} 0 & \text{if } s(\mathrm{m}, \theta) < r(\mathrm{m}, \theta), \\ b(m_r, m_s) & \text{if } r(\mathrm{m}, \theta) < s(\mathrm{m}, \theta). \end{cases}$$

Further, if $1 \leq r < s \leq q$ are such that $s(\mathrm{m}, \theta) < r(\mathrm{m}, \theta)$, then

$$m_s = m_{s(\mathrm{m},\theta)} \leq m_{r(\mathrm{m},\theta)} = m_r \implies m_s = m_{s(\mathrm{m},\theta)} = m_{r(\mathrm{m},\theta)} = m_r$$

and hence (2) assures that $b(m_r, m_s)$ is even. So, m-excellence property (2) implies

$$f_\theta(0, T, x) = \prod_{1 \leq r < s \leq q} (t_r - t_s)^{b(m_r, m_s)}.$$

Secondly, suppose the m-excellence property (3) holds. Then, it allows us to infer the existence of a nonzero homogeneous (in $T$) polynomial $g_\theta \in \mathbb{Q}[T]$ such that $f_\theta(0, T, x) = g_\theta^2$. In either case, note that $f_\theta(0, T, x)$ is homogeneous in $T$ of $T$-degree $d(\mathrm{m})$. So, (ii) holds. Lastly, if $\varepsilon$ satisfies m-excellence property (2), then in view of the above,

$$f(0, T, x) = |G(\mathfrak{m})| \cdot \prod_{1 \leq r < s \leq q} (t_r - t_s)^{b(m_r, m_s)}$$

is indeed a nonzero $T$-homogeneous polynomial of $T$-degree $d(\mathfrak{m})$. On the other hand, if $\varepsilon$ satisfies $\mathfrak{m}$-excellence property (3), then as observed above, we have

$$f(0, T, x) = \sum_{\theta \in G(\mathfrak{m})} g_\theta^2,$$

which is nonzero by (i) of Theorem 7 and obviously it is a $T$-homogeneous polynomial of $T$-degree $d(\mathfrak{m})$. This proves (iii). $\qquad \square$

Now we are ready to state and prove one of our main results. The following theorem is a slightly stronger version of Theorem 3 of [10] with a more streamlined proof.

**Theorem 8** *We continue to employ the above notation. In particular, let* $\mathfrak{m} : 0 = m_0 < m_1 \leq \cdots \leq m_q$ *be as above. Assume that* $q \geq 2$.

*(i) Suppose* $\varepsilon$ *is excellent. Then,*

$$Symm_N(v(z, \pi[N], \varepsilon)) \neq 0.$$

*(ii) Suppose* $2a, c, p$ *are positive integers (*$a$ *is allowed to be a half-integer which need not be an integer) and* $\mathfrak{n} : 0 = n_0 < n_1 < \cdots < n_p = N$ *is a sequence of integers such that for each positive integer* $r$,

$$2 \cdot a \cdot |\{i \mid 1 \leq i \leq p, \quad n_i - n_{i-1} = r\}| \quad \text{is an even integer.}$$

*Let* $M := M(\mathfrak{n}, a, c)$ *and* $W := (w_1, \ldots, w_N)$, *where* $w_i := w_i(\mathfrak{n})$ *for* $1 \leq i \leq N$. *Then, we have* $M \in E(N, W)$ *and* $Symm_N(\delta(z, M)) \neq 0$.

*Proof* Let $\mu := v(z, \pi[N], \varepsilon)$. In view of the excellence property (1) of $\varepsilon$, we have $\mu = v(z, \pi[\mathfrak{m}], \varepsilon)$. For $\theta \in S_N \setminus G(\mathfrak{m})$, define $U_\theta(\mathfrak{m}, \varepsilon) := \phi(\theta(\mu))$ and let

$$U(\mathfrak{m}, \varepsilon) := \sum_{\theta \in S_N \setminus G(\mathfrak{m})} U_\theta(\mathfrak{m}, \varepsilon).$$

Then, it is clear that

$$\phi(Symm_N(\mu)) = V(\mathfrak{m}, \varepsilon) + U(\mathfrak{m}, \varepsilon).$$

From (iv) of Lemma 2 it follows that for each $\theta \in S_N \setminus G(\mathfrak{m})$, the $t$-order of $U_\theta(\mathfrak{m}, \varepsilon)$ is positive, i.e., the polynomial $U_\theta(\mathfrak{m}, \varepsilon)$ is divisible by $t$ in the ring $k[t, T, x]$. Hence the $t$-order of $U(\mathfrak{m}, \varepsilon)$ is also positive. On the other hand, by (iii) of Lemma 3, the $t$-order of $V(\mathfrak{m}, \varepsilon)$ is 0, i.e., the polynomial $V(\mathfrak{m}, \varepsilon)$ is not divisible by $t$ in the ring

$k[t, T, x]$. Consequently, $\phi(Symm_N(\mu)) \neq 0$. But then, $Symm_N(\mu)$ must be nonzero. Thus assertion (i) holds.

Now let $\mathfrak{n}$ and $M$ be as in (ii). It is straightforward to verify that $M$ belongs to $E(N, W)$. Next, let $q := \max\{n_i - n_{i-1} \mid 1 \leq i \leq p\}$ and for $1 \leq r \leq q$, let

$$\Gamma(r) := \{1 \leq i \leq p \mid (n_i - n_{i-1}) \geq r\} \quad \text{and} \quad J_r := \{r + n_{i-1} \mid i \in \Gamma(r)\}.$$

Let $\gamma(r)$ denote the cardinality of $\Gamma(r)$. Obviously, $J_r$ has cardinality $\gamma(r) \geq 1$ for $1 \leq r \leq q$, and the sets $J_1, \ldots, J_q$ form a partition of $\{1, \ldots, N\}$. Let $e(i, j)$ denote the $(i, j)$th entry of $M$. Then $e(i, j) = 0$ if and only if $(i, j) \in J_r \times J_r$ for some $r$ with $1 \leq r \leq q$. Also, $e(i, j) = 2a$ if and only if $(i, j) \in J_r \times J_s$, for some $(r, s)$ with $(i - r) = (j - s)$ and $1 \leq r \neq s \leq q$. Moreover, it can be easily verified that

$$\sum_{(i,j) \in J_r \times J_s} e(i, j) = c\gamma(r)\gamma(s) + (2a - c) \cdot \min\{\gamma(r), \gamma(s)\}.$$

Suppose $\tau$ is a permutation of $\{1, \ldots, q\}$ such that $\gamma(\tau(i)) \leq \gamma(\tau(j))$ for $1 \leq i < j \leq q$. Define $\mathfrak{m} : 0 = m_0 < m_1 \leq \cdots \leq m_q$ by setting $m_i := \gamma(\tau(i))$ for $1 \leq i \leq q$. Then clearly $m_1 + m_2 + \cdots + m_q = N$. Observe that by hypothesis, $2am_r$ is an even integer for $1 \leq r \leq q$. Since $J_{\tau(1)}, J_{\tau(2)}, \ldots, J_{\tau(q)}$ clearly form a partition of $\{1, 2, \ldots, N\}$, there exists $\theta \in S_N$ such that

$$A_r(\mathfrak{m}) := \left\{\theta(i) \mid i \in J_{\tau(r)}\right\} = \left\{i + \sum_{j=0}^{r-1} m_j \,\middle|\, 1 \leq i \leq m_r\right\}.$$

Fix such a permutation $\theta$. Observe that $q = 1$ if and only if $N = p$ if and only if $M = 0$. Since $Symm_N(\delta(z, 0)) = N! \neq 0$, we henceforth assume $q \geq 2$. Define $\varepsilon : \pi[N] \to \mathbb{N}$ by

$$\varepsilon(i, j) := e(\theta^{-1}(i), \theta^{-1}(j)) \quad \text{for } (i, j) \in \pi[N].$$

Then, $\varepsilon(i, j) = 0$ if and only if $(i, j) \in \pi(A_r(\mathfrak{m}))$ for $1 \leq r \leq q$, i.e., $\varepsilon$ satisfies the $\mathfrak{m}$-excellence property (1). Moreover, we have

$$\theta(\delta(z, M)) = \prod_{(i,j) \in \pi[N]} (z_{\theta(i)} - z_{\theta(j)})^{e(i,j)} = \pm \prod_{(i,j) \in \pi[N]} (z_i - z_j)^{\varepsilon(i,j)}.$$

Now for $1 \leq r < s \leq q$, we have

$$\sum_{(i,j) \in A_r(\mathfrak{m}) \times A_s(\mathfrak{m})} \varepsilon(i, j) = \sum_{(i,j) \in J_{\tau(r)} \times J_{\tau(s)}} e(i, j)$$

and hence letting $b(m_r, m_s) := cm_rm_s + (2a - c)m_r$, it follows that

$$\sum_{(i,j)\in A_r(\mathrm{m})\times A_s(\mathrm{m})} \varepsilon(i,j) = b(m_r, m_s).$$

Furthermore, $(m_r, m_r) = cm_r(m_r - 1) + 2am_r$ is clearly an even integer. Thus $\varepsilon$ is m-excellent. As already observed,

$$Symm_N(\delta(z, M)) = Symm_N(\theta(\delta(z, M)) = \pm Symm_N(v(z, \pi[N], \varepsilon)).$$

Since $Symm_N(v(z, \pi[N], \varepsilon)) \neq 0$ by assertion (i), assertion (ii) holds. $\qquad\square$

**Corollary** *Let $m, n$ be positive integers such that $m \geq 2$ and $N = mn$. For $1 \leq r < s \leq m$, define*

$$T(r, s) := \{\{l_1 m + r,\ l_2 m + s\} \mid 0 \leq l_1, l_2 \leq n - 1\}.$$

*As declared before, we identify $T(r, s)$ as a subset of $\pi[N]$.*

*(i) Let $\alpha : \pi[N] \to \mathbb{N}$ be a function such that*

$$\alpha(i, j) = 0 \quad \text{if and only if } i \equiv j \bmod m,$$

*and for $1 \leq r < s \leq m$,*

$$\sum_{(i,j)\in T(r,s)} \alpha(i, j) \quad \text{is an even integer.}$$

*Then, we have*

$$Symm_N(v(z, \pi[N], \alpha)) \neq 0.$$

*(ii) Let $2a, c$ be positive integers with $2an$ even. Then,*

$$M := M(m, n, a, c) \in E(N,\ (2a + cn - c)(m - 1))$$

*and $Symm_N(\delta(z, M)) \neq 0$.*

*Proof* For the proof of (i), the reader is referred to (i) of the Corollary to Theorem 3 of [10]. Since the hypothesis of (ii) differs from the hypothesis of (ii) of the Corollary to Theorem 3 of [10], we proceed to prove (ii). Let $M$ be as in (ii). Define

$$\mathfrak{n} := 0 < m < 2m < \cdots < im < \cdots < nm = N.$$

Now it is straightforward to verify that $M = M(\mathfrak{n}, a, c)$. Either using the block-format description of $M(m, n, a, c)$ presented in the remarks following the definition of $M(m, n, a, c)$, or by (ii) of Theorem 8, we see that $M$ is in $E(N, (2a + cn - c)(m - 1))$. Given an integer $r$, the set $\{i \mid 1 \leq i \leq n, im - (i - 1)m = r\}$ is empty if $r \neq m$ and $\{1, 2, \ldots, n\}$ if $r = m$. Since $0 = 2a|\emptyset|$ is of course an even integer and $2an$ is assumed to be an even integer, the hypotheses of (ii) of Theorem 8 are satisfied. So, assertion (ii) follows from (ii) of Theorem 8. Alternatively, (ii) can be derived from (i) (the details are left to the reader).                                             $\square$

*Examples* Consider the $6 \times 6$ block-matrices

$$E_1 := \begin{bmatrix} 0 & C_1 \\ C_1^T & 0 \end{bmatrix} \quad \text{and} \quad E_2 := \begin{bmatrix} 0 & C_2 \\ C_2^T & 0 \end{bmatrix},$$

where

$$C_1 := \begin{bmatrix} 3 & 3 & 3 \\ 3 & 4 & 3 \\ 3 & 3 & 4 \end{bmatrix} \quad \text{and} \quad C_2 := \begin{bmatrix} 3 & 3 & 3 \\ 3 & 3 & 4 \\ 3 & 3 & 4 \end{bmatrix}.$$

Then, brute force computation shows that

$$Symm_6(\delta(z, E_1)) = 0 \quad \text{and} \quad Symm_6(\delta(z, E_2)) \neq 0.$$

Any satisfactory generalization of Theorem 8 should be able to differentiate between $E_1$ and $E_2$.

As before, let us consider an $N \times N$ symmetric matrix with nonnegative integer entries which is expressible as a $q \times q$ ($q$, a positive integer) block-matrix $[C_{ij}]$, where the size of $C_{rr}$ is $m_r \times m_r$ and the block-sizes are nondecreasing positive integers, i.e., $1 \leq m_r \leq m_{r+1}$ for $1 \leq r < q$. Let $\varepsilon : \pi[N] \to \mathbb{N}$ be the function associated with $[C_{ij}]$ and let $\mathrm{m} : 0 = m_0 < m_1 \leq \cdots \leq m_q$. Let $A_1(\mathrm{m}), \ldots, A_q(\mathrm{m})$ and $\pi[\mathrm{m}]$ be the sets as above. Of course, each diagonal block $C_{rr}$ is itself symmetric. Henceforth, we assume that the diagonal entries of each $C_{rr}$ are all 0. Clearly, each $C_{rr}$ is determined by the restriction of $\varepsilon$ to $\pi(A_r(\mathrm{m}))$. Let $G_r(\mathrm{m})$ be the group of permutations of the set $A_r(\mathrm{m})$. Then, we have the following nonvanishing theorem (see [11]). Since this theorem is used (in an essential way) only in the proof of assertion (iv) of Theorem 12, we have chosen to present an example instead of a formal proof.

**Theorem 9** *Let the notation be as above. Assume that the following holds.*

*(i)* $m_1 < m_2 < \cdots < m_q$.
*(ii)* *For* $1 \leq r \leq q$, *we have*

$$\sum_{\theta \in G_r(\mathrm{m})} \theta(v(z, \pi(A_r(\mathrm{m})), \varepsilon)) \neq 0.$$

*(iii)* *For* $1 \leq r \leq q$,

$$max\{\varepsilon(i,j) \mid (i,j) \in \pi(A_r(\mathrm{m}))\} < min\{\varepsilon(i,j) \mid (i,j) \in \pi[\mathrm{m}]\}.$$

*Then,* $Symm_N(v(z, \pi[N], \varepsilon)) \neq 0$.

*Example* Let $N = 3$ and $\mathrm{m} : 0 < 1 < 2$. In this case $q = 2$, $A_1(\mathrm{m}) = \{1\}$ and $A_2(\mathrm{m}) = \{2, 3\}$. Clearly, $\pi(A_1(\mathrm{m})) = \emptyset$, $\pi(A_2(\mathrm{m})) = \{(2, 3)\}$, $G_1(\mathrm{m})$ has only the identity permutation and $G_2(\mathrm{m})$ consists of the identity permutation together with the transposition $(2, 3)$. Let $\mu := v(z, \pi[3], \varepsilon)$. Then,

$$\mu = (z_1 - z_2)^{\varepsilon(1,2)} (z_1 - z_3)^{\varepsilon(1,3)} (z_2 - z_3)^{\varepsilon(2,3)}.$$

Let $\mu_r := v(z, \pi(A_r(\mathrm{m}), \varepsilon)$ for $r = 1, 2$. Since an empty product is 1, we have $\mu_1 = 1$. Of course, $\mu_2 = (z_2 - z_3)^{\varepsilon(2,3)}$. The $G_1(\mathrm{m})$-symmetrization of $\mu_1$ is 1 (whence nonzero). The $G_2(\mathrm{m})$-symmetrization of $\mu_2$ is the polynomial $(z_2 - z_3)^{\varepsilon(2,3)} + (z_3 - z_2)^{\varepsilon(2,3)}$, which is nonzero if and only if $\varepsilon(2, 3)$ is an even integer. Henceforth, $\varepsilon(2, 3)$ is tacitly assumed to be an even integer. Note that $\pi[\mathrm{m}] = \{(1, 2), (1, 3)\}$. Consequently, $G(\mathrm{m}) = G_2(\mathrm{m})$. It is more fruitful to regard $G(\mathrm{m})$ as the internal direct product of its subgroups $G_1(\mathrm{m})$ and $G_2(\mathrm{m})$. Consider $\theta \in S_3$ and then substitute $z_1 = tx_1 + t_1, z_2 = tx_2 + t_2, z_3 = tx_3 + t_2$ in $\theta(\mu)$. Firstly, suppose $\theta \in G(\mathrm{m})$. Then, the substituted $\theta(\mu)$ is one of the following two polynomials:

$$t^{\varepsilon(2,3)} (x_2 - x_3)^{\varepsilon(2,3)} [t(x_1 - x_2) + (t_1 - t_2)]^{\varepsilon(1,2)} [t(x_1 - x_3) + (t_1 - t_2)]^{\varepsilon(1,3)},$$
$$t^{\varepsilon(2,3)} (x_3 - x_2)^{\varepsilon(2,3)} [t(x_1 - x_3) + (t_1 - t_2)]^{\varepsilon(1,2)} [t(x_1 - x_2) + (t_1 - t_2)]^{\varepsilon(1,3)}.$$

Denote these by $\alpha_1$, $\alpha_2$ respectively. Let $\alpha$ be defined by: $\alpha_1 + \alpha_2 = t^{\varepsilon(2,3)} \cdot \alpha$. Then, substituting $t = 0$ in $\alpha$ yields

$$(t_1 - t_2)^{\varepsilon(1,2)+\varepsilon(1,3)} \cdot \left[ (x_2 - x_3)^{\varepsilon(2,3)} + (x_3 - x_2)^{\varepsilon(2,3)} \right].$$

Nonzero-ness of the above polynomial, which is equivalent to nonzero-ness of the $G_2(\mathrm{m})$-symmetrization of $\mu_2$, allows us to infer that the $t$-order of $\alpha_1 + \alpha_2$ is $\varepsilon(2, 3)$. Next, suppose $\theta \in S_3 \setminus G(\mathrm{m})$. Then, the substituted $\theta(\mu)$ is one of

$$t^{\varepsilon(1,3)} (x_2 - x_3)^{\varepsilon(1,3)} [t(x_2 - x_1) + (t_2 - t_1)]^{\varepsilon(1,2)} [t(x_1 - x_3) + (t_1 - t_2)]^{\varepsilon(2,3)},$$
$$t^{\varepsilon(1,2)} (x_3 - x_2)^{\varepsilon(1,2)} [t(x_3 - x_1) + (t_2 - t_1)]^{\varepsilon(1,3)} [t(x_2 - x_1) + (t_2 - t_1)]^{\varepsilon(2,3)},$$
$$t^{\varepsilon(1,2)} (x_2 - x_3)^{\varepsilon(1,2)} [t(x_2 - x_1) + (t_2 - t_1)]^{\varepsilon(1,3)} [t(x_3 - x_1) + (t_2 - t_1)]^{\varepsilon(2,3)},$$
$$t^{\varepsilon(1,3)} (x_3 - x_2)^{\varepsilon(1,3)} [t(x_3 - x_1) + (t_2 - t_1)]^{\varepsilon(1,2)} [t(x_1 - x_2) + (t_1 - t_2)]^{\varepsilon(2,3)}.$$

Name these polynomials $\beta_1$, $\beta_2$, $\beta_3$, $\beta_4$ respectively. Then, observe that the $t$-order of $\beta_1 + \beta_2 + \beta_3 + \beta_4$ is at least $\min\{\varepsilon(1, 2), \varepsilon(1, 3)\}$. At this point, the hypothesis (iii) of Theorem 9 ensures that $\varepsilon(2, 3) < \min\{\varepsilon(1, 2), \varepsilon(1, 3)\}$, i.e., the sum $\alpha_1 + \alpha_2 + \beta_1 + \beta_2 + \beta_3 + \beta_4$ has $t$-order $\varepsilon(2, 3)$. Thus, $Symm_3(\mu)$ has to be a nonzero polynomial. Importantly, we observe that Theorem 9 does offer something different from Theorem 8 even in this case of $N = 3$. Suppose each of $\varepsilon(1, 3)$, $\varepsilon(2, 3)$ is a strictly positive even integer, $\varepsilon(1, 2)$ is a strictly positive odd integer and $\varepsilon(1, 2) > \varepsilon(2, 3)$ as well as $\varepsilon(1, 3) > \varepsilon(2, 3)$. We have only two choices for blocking-sequences: either $0 < 1 < 2$ or $0 < 1 \leq 1 \leq 1$. Note that in neither case the excellence properties are satisfied by our $\varepsilon$.

So far, we have focused our attention on symmetrizations of $\prod(z_i - z_j)^{\varepsilon(i,j)}$, where the exponents $\varepsilon(i,j)$ were assumed to be nonnegative integers. In the following two theorems, we prove nonvanishing of the symmetrizations of products of the form $\prod(z_i - z_j)^{\varepsilon(i,j)}$, where each $\varepsilon(i,j)$ is a nonpositive integer and the function $\varepsilon$ satisfies certain other conditions. These theorems provide us tools for constructing semi-invariants of the kind that can not be readily constructed using Theorems 7–9. As to be expected, we need some new definitions, terminology and notation.

**Definitions** Let $m$, $n$ be positive integers and let $A := [a(i,j)]$ be an $m \times n$ matrix with nonnegative integer entries. Let $T_1, \ldots, T_m$ be indeterminates and let $T$ stand for $(T_1, \ldots, T_m)$.

1. By $\max(A)$, we mean $\max\{a(i,j) \mid 1 \le i \le m, \ \ 1 \le j \le n\}$.
2. For $1 \le r \le m$, define

$$co(r, A) := \{j \mid 1 \le j \le n \text{ and } a(r,j) = \max(A)\}$$

and let $|co(r, A)|$ denote the cardinality of $co(r, A)$. Let

$$co(A) := \bigcup_{r=1}^{m} co(r, A).$$

3. For $1 \le r \le m$, define

$$sp(r, A) := \{j \mid 1 \le j \le n \text{ and } a(r,j) \ne 0\}$$

and let $|sp(r, A)|$ denote the cardinality of $sp(r, A)$. Let

$$sp(A) := \bigcup_{r=1}^{m} sp(r, A).$$

4. For $1 \le r \le m$, define

$$b(r, A) := \sum_{j \in sp(r,A) \setminus co(A)} a(r, j).$$

5. For $1 \le r < s \le m$, define

$$\nu(r, s, A) := \sum_{j \in co(s,A)} a(r, j) + \sum_{j \in co(r,A)} a(s, j).$$

6. Define

$$pol(A, T) := \prod_{r=1}^{m} T_r^{b(r,A)} \prod_{1 \le r < s \le m} (T_r - T_s)^{\nu(r,s,A)}.$$

7. As usual, let $S_m$ denote the permutation group of $\{1, \ldots, m\}$. Given a polynomial $f(T_1, \ldots, T_m) \in \mathbb{Q}[T_1, \ldots, T_m]$ and a permutation $\theta \in S_m$, by $\theta(f(T))$, we mean the polynomial $f(T_{\theta(1)}, \ldots, T_{\theta(m)})$. Define

$$grp(A) := \{\theta \in S_m \mid |co(r, A)| = |co(\theta(r), A)| \quad \text{for } 1 \leq r \leq m\}$$

and set

$$rat(A, T) := \{\theta(pol(A, T))^{-1} \mid \theta \in grp(A)\}.$$

8. For an $r \times s$ matrix $A := [a_{ij}]$, define the *norm* of $A$ to be

$$\|A\| := \sum_{j=1}^{s} \sum_{i=1}^{r} a_{ij}.$$

*Example* Explicitly, let us consider the $3 \times 5$ matrix

$$A := \begin{bmatrix} 0 & 2 & 1 & 0 & 2 \\ 2 & 1 & 0 & 2 & 1 \\ 1 & 0 & 2 & 1 & 0 \end{bmatrix}.$$

Then, since $max(A)$ stands for the greatest entry of $A$, we have $max(A) = 2$. Observe that

$$co(1, A) = \{2, 5\}, \quad co(2, A) = \{1, 4\} \quad co(3, A) = \{3\}$$

and hence $co(A) = \{1, 2, 3, 4, 5\}$. By its very definition, $grp(A)$ consists of the permutations of the rows of $A$ that respect the cardinalities of the sets $co(1, A)$, $co(2, A)$, and $co(3, A)$. Obviously, each permutation in $grp(A)$ must fix 3 and leave the set $\{1, 2\}$ invariant. Thus

$$grp(A) = \{id, \tau\} < S_3,$$

where $id$ is the identity permutation and $\tau$ is the transposition $(1, 2)$. Since $sp(r, A) \subset \{1, 2, 3, 4, 5\}$, the quantity $b(r, A)$ is an empty sum for $r = 1, 2, 3$. So, $b(r, A) = 0$ for $r = 1, 2, 3$. Note that

$$\nu(1, 2, A) = a(1, 1) + a(1, 4) + a(2, 2) + a(2, 5) = 2,$$
$$\nu(1, 3, A) = a(1, 3) + a(3, 2) + a(3, 5) = 1,$$
$$\nu(2, 3, A) = a(2, 3) + a(3, 1) + a(3, 4) = 2.$$

Now, by its definition, the polynomial $pol(A, T)$ associated with $A$ is given by

$$pol(A, T) = (T_1 - T_2)^2(T_1 - T_3)(T_2 - T_3)^2.$$

The set $rat(A, T)$ consists of reciprocals of $pol(A, T_{\theta(1)}, T_{\theta(2)}, T_{\theta(3)})$ as $\theta$ ranges over the permutations in $grp(A)$. Thus $rat(A, T)$ is the set

$$\left\{ \frac{1}{(T_1 - T_2)^2 (T_1 - T_3)(T_2 - T_3)^2}, \quad \frac{1}{(T_2 - T_1)^2 (T_2 - T_3)(T_1 - T_3)^2} \right\}.$$

The above introduced definitions and terminology equip us to define a criterion which plays a key role in the formulation as well as the proof of the following Theorem 10.

**Definition**   Suppose $m$, $n$ are positive integers and $A := [a(i, j)]$ is an $m \times n$ matrix, where each entry $a(i, j)$ is a nonnegative integer. Then, $A$ is said to be *admissible* if the following three requirements are satisfied.

(1)  $co(r, A) \neq \emptyset$ for $1 \leq r \leq m$ and

$$co(r, A) \cap co(s, A) = \emptyset \quad \text{for } 1 \leq r < s \leq m.$$

In other words, at least one entry in each row of $A$ equals $max(A)$ and each column of $A$ has at most one (possibly none) entry equal to $max(A)$.
(2)  $rat(A, T)$ is a $\mathbb{Q}$-linearly independent set of rational functions.
(3)  For each pair $(M, r)$, such that $M$ is a $p \times q$ submatrix of $A$ with $p, q \geq 2$ and $p + q - 1 = |co(r, A)|$, we have $\|M\| < (p + q - 1)max(A)$.

*Remarks*

1.  Let $A$ be an $m \times n$ matrix with nonnegative integer entries. If $m = 1$, then it is easy to verify that $A$ is admissible. On the other hand, if $m \geq n + 1$, then either some row of $A$ does not have $max(A)$ as an entry or $max(A)$ occurs in at least two distinct rows of some column of $A$; in either case, $A$ is not admissible. Note that the admissibility condition is not symmetric in rows and columns; in fact, even when $m = n$, admissibility of $A$ need not guarantee admissibility of $A^T$ (an explicit example is left to the reader to construct).
2.  Let $A$ be an $m \times n$ matrix with nonnegative integer entries such that $A$ satisfies the requirement (1) in the definition of admissibility. Now if $|co(r, A)| \leq 2$ for $1 \leq r \leq m$, then the requirement (3) holds vacuously ($p + q - 1 \geq 3$ implies $p + q - 1 \neq |co(r, A)|$ for all $r$).
3.  Let $A$ be an $m \times n$ matrix with nonnegative integer entries satisfying requirements (1) and (3) in the definition of admissibility. If $|co(r, A)| \neq |co(s, A)|$ for $1 \leq r < s \leq m$, then $grp(A)$ is the trivial group and hence $rat(A, T)$ contains only one nonzero rational function. It follows that $rat(A, T)$ is $\mathbb{Q}$-linearly independent and thus $A$ is admissible.
4.  Let $A$ be an $m \times n$ matrix with nonnegative integer entries satisfying requirements (1) and (3) in the definition of admissibility. If $pol(A, T)$ is fixed by each permutation of $grp(A)$, then $rat(A, T)$ contains only one nonzero rational function and hence $A$ is admissible.

5. Let $A$ be an $m \times n$ matrix with nonnegative integer entries satisfying requirements (1) and (3) in the definition of admissibility and such that $grp(A) = \{id, \theta\}$
   with $\theta \neq id$. In this case, observe that $A$ fails to be admissible if and only if
   $\theta(pol(A, T)) = (-1) \cdot pol(A, T)$.
6. Let $A$ be an $m \times n$ matrix with nonnegative integer entries satisfying requirements
   (1) and (3) in the definition of admissibility and such that $\nu(r, s, A) = 0$ for $1 \leq$
   $r < s \leq m$. Then, $pol(A, T)$ is the power-product $T_1^{b(1,A)} \cdots T_m^{b(m,A)}$. Since any
   subset of power-products of indeterminates $T_1, \ldots, T_m$ is $\mathbb{Q}$-linearly independent,
   $rat(A, T)$ is $\mathbb{Q}$-linearly independent. Thus $A$ is admissible.

*Examples*

1. Let $A$ be the $3 \times 5$ matrix in the example just above the definition of admissibility.
   Then, it is clear that $A$ satisfies the requirement (1) of admissibility. Since $rat(A, T)$
   is the set

   $$\left\{ \frac{1}{(T_1 - T_2)^2(T_1 - T_3)(T_2 - T_3)^2}, \quad \frac{1}{(T_2 - T_1)^2(T_2 - T_3)(T_1 - T_3)^2} \right\}$$

   and $(T_1 - T_3)$, $(T_2 - T_3)$ are readily seen to be $\mathbb{Q}$-linearly independent, $rat(A, T)$
   is $\mathbb{Q}$-linearly independent. Since $|co(r, A)| \leq 2$ for $r = 1, 2, 3$, the requirement (3)
   in the definition of admissibility is vacuously satisfied. Therefore, $A$ is admissible.
2. Consider the $3 \times 3$ matrix

   $$A := \begin{bmatrix} 2 & 1 & 1 \\ 0 & 2 & 1 \\ 0 & 0 & 2 \end{bmatrix}.$$

   Clearly, $max(A) = 2$, $co(1, A) = \{1\}$, $co(2, A) = \{2\}$ and $co(3, A) = \{3\}$. Hence
   $co(A) = \{1, 2, 3\}$ and $grp(A) = S_3$. As in the above example, we have $b(r, A) = 0$
   (an empty sum) for $r = 1, 2, 3$. Next, observe that

   $$\begin{aligned} \nu(1, 2, A) &= a(1, 2) + a(2, 1) = 1, \\ \nu(1, 3, A) &= a(1, 3) + a(3, 1) = 1, \\ \nu(2, 3, A) &= a(2, 3) + a(3, 2) = 1. \end{aligned}$$

   Hence $pol(A, T) = (T_1 - T_2)(T_1 - T_3)(T_2 - T_3)$. It follows that

   $$rat(A, T) = \left\{ \frac{1}{pol(A, T)}, \frac{-1}{pol(A, T)} \right\}.$$

   Obviously, $rat(A, T)$ is a $\mathbb{Q}$-linearly dependent set. So, $A$ is *not* admissible.
3. Suppose $m, n, d$ are positive integers and some nonempty sets $C_1, \ldots, C_m$ form a
   partition of $\{1, \ldots, n\}$. Let $A := [a(i, j)]$ be an $m \times n$ matrix with each $a(i, j)$ being
   a nonnegative integer such that $a(i, j) = 0$ if $j \in \{1, \ldots, n\} \setminus C_i$ for $1 \leq i \leq m$ and
   the maximum of the set $\{a(i, j) \mid j \in C_i\}$ is $d$. Then $max(A) = d$ and we have
   $\emptyset \neq co(r, A) \subseteq C_r$ for $1 \leq r \leq m$. In particular, $co(r, A) \cap co(s, A) = \emptyset$ for $1 \leq$

$r < s \leq m$. Note that $\nu(r, s, A) = 0$ for $1 \leq r < s \leq m$ and hence (as remarked above) $rat(A, T)$ is a $\mathbb{Q}$-linearly independent set. To verify the condition (3) in the definition of admissibility, suppose $M$ is a $p \times q$ submatrix of $A$ such that $p \geq 2$, $q \geq 2$ and suppose $r$ is a positive integer $\leq m$ with $p + q - 1 = |co(r, A)|$. Since each column of our matrix $A$ contains at most one nonzero entry, each column of $M$ contains at most one nonzero entry. Furthermore, since each nonzero entry of $A$ is at most $d$, each nonzero entry of $M$ is at most $d$. Consequently, $\|M\| \leq qd$. Now $p \geq 2$ implies $qd < (p + q - 1)d$. Thus $A$ is indeed admissible.

**Definitions** Let $m$, $N$ be integers with $1 \leq m \leq N - 2$. Let $t$, $y$, $T_1, \ldots, T_m$ and $x_1, \ldots, x_N$ and $z_1, \ldots, z_N$ be indeterminates and let $x$ stand for $(x_1, \ldots, x_N)$, $T$ stand for $(T_1, \ldots, T_m)$ and $z$ stand for $(z_1, \ldots, z_N)$.

1. Let $A := [a(i, j)]$ be an $m \times (N - m)$ matrix with nonnegative integer entries $a(i, j)$.

2. Let $E(A) \in E(N)$ be the matrix given in block-form by

$$E(A) := \begin{bmatrix} 0 & A \\ A^T & 0 \end{bmatrix}.$$

3. For $1 \leq r \leq m$, let

$$C(r, A) := \{r\} \cup \{m + j \mid j \in co(r, A)\} \quad \text{and}$$
$$J(r, A) := \{r\} \cup \{m + j \mid j \in sp(r, A)\}.$$

4. Define

$$C(A) := C(1, A) \cup \cdots \cdots \cup C(m, A) \quad \text{and}$$
$$J(A) := J(1, A) \cup \cdots \cdots \cup J(m, A).$$

5. For $\theta \in S_N$, let

$$C^{\theta}(A) := \{\theta(C(1, A)), \ldots \ldots, \theta(C(m, A))\}.$$

6. Let $\psi : k[z] \to k[t, y, T, x, z]$ be the $k$-homomorphism of rings such that for $1 \leq i \leq N$,

$$\psi(z_i) := \begin{cases} tx_i + T_r & \text{if } i \in C(r, A) \text{ with } 1 \leq r \leq m, \\ yx_i & \text{if } i \in J(A) \setminus C(A), \\ z_i & \text{otherwise.} \end{cases}$$

7. Let $id \in S_N$ denote the identity permutation of $\{1, \ldots, N\}$. Define

$$G(A) := \{\theta \in S_N \mid \theta(J(A)) = J(A)\},$$
$$H(A) := \{\theta \in G(A) \mid C^{\theta}(A) = C^{id}(A)\}.$$

8. A permutation $\tau$ of $\{1, \ldots, m\}$ is said to be *induced by* $\theta \in S_N$ provided we have $\theta(C(r, A)) = C(\tau(r), A)$ for $1 \le r \le m$.

**Lemma 4** *Let the notation be as above. Suppose $co(r, A) \ne \emptyset$ for $1 \le r \le m$ and*

$$co(r, A) \cap co(s, A) = \emptyset \quad \text{for } 1 \le r < s \le m.$$

*Then, the following holds.*

(i)  $C(r, A) \cap C(s, A) = \emptyset$ *for* $1 \le r < s \le m$.
(ii)  *$\psi$ is a well-defined injective homomorphism of rings.*
(iii)  *An element $\theta \in H(A)$ induces a unique permutation $\tau$ of $\{1, \ldots, m\}$; further-more, $\tau \in grp(A)$.*
(iv)  *The $z$-degree of $\psi(\theta(\delta(z, E(A)))$ is $\ge 1$ if and only if $\theta \in S_N \setminus G(A)$.*

*Proof* Assertion (i) follows from the hypothesis. In view of (i), the map $\psi$ is at once seen to be well-defined. Obviously, $\psi$ is an injective $k$-homomorphism of rings as asserted in (ii). To prove (iii), fix $\theta \in H(A)$. By the very definition of $H(A)$, $\theta$ deter-mines a unique permutation $\tau$ of $\{1, \ldots, m\}$ such that $\theta(C(r, A)) = C(\tau(r), A)$ for $1 \le r \le m$. In particular, $|C(r, A)| = |C(\tau(r), A)|$ for $1 \le r \le m$. Now $|C(r, A)| = 1 + |co(r, A)|$ for $1 \le r \le m$ and hence $|co(r, A)| = |co(\tau(r), A)|$ for $1 \le r \le m$, i.e., $\tau \in grp(A)$.

Lastly, we prove (iv). Define $\mu(z) := \delta(z, E(A))$. Given $\theta \in G(A)$, $1 \le i \le m$ and $j \in J(A) \setminus \{1, \ldots, m\}$, observe that $\psi(z_{\theta(i)} - z_{\theta(j)}) =$

$$\begin{cases} t(x_{\theta(i)} - x_{\theta(j)}) + T_r - T_s & \text{if } \theta(i) \in C(r, A) \text{ and } \theta(j) \in C(s, A), \\ tx_{\theta(i)} + T_r - yx_{\theta(j)} & \text{if } \theta(i) \in C(r, A) \text{ and } \theta(j) \in J(A) \setminus C(A), \\ yx_{\theta(i)} - tx_{\theta(j)} - T_r & \text{if } \theta(i) \in J(A) \setminus C(A) \text{ and } \theta(j) \in C(r, A), \\ yx_{\theta(i)} - yx_{\theta(j)} & \text{if } \theta(i) \in J(A) \setminus C(A) \text{ and } \theta(j) \in J(A) \setminus C(A). \end{cases}$$

Thus for $\theta \in G(A)$, the (total) $z$-degree of $\psi(\theta(\mu(z)))$ is 0. Suppose $\theta \in S_N \setminus G(A)$ and fix $s \in J(A)$ such that $\theta(s)$ is not in $J(A)$. If $1 \le s \le m$, then for any $j \in sp(s, A)$, the $z$-degree of $\psi(z_{\theta(s)} - z_{\theta(m+j)})$ is 1. Also, if $m < s$, then for an $r \in \{1, \ldots, m\}$ such that $s \in J(r, A)$, the $z$-degree of $\psi(z_{\theta(r)} - z_{\theta(s)})$ is 1. Therefore, the $z$-degree of $\psi(\theta(\mu(z)))$ is $\ge 1$ if and only if $\theta \in S_N \setminus G(A)$. $\square$

*Remark* In view of (ii) of the above lemma, $\psi$ naturally extends to an injective $k$-homomorphism $k(z) \to k(t, y, T, x, z)$ of fields. For economy of notation, this field-homomorphism is also denoted by $\psi$.

**Definitions** We continue to use the above notation. Recall that $A := [a_{ij}]$ is an $m \times (N - m)$ matrix with nonnegative integer entries $a(i, j)$.

1. For $1 \le r \le m$, define

$$\alpha(r, A) := \prod_{j \in co(r, A)} (z_r - z_{m+j}) \quad \text{and} \quad \beta(r, A) := \prod_{j \in J_r(A) \setminus C(A)} (z_r - z_j)^{a(r, j)}.$$

2. For $1 \leq r < s \leq m$, define

$$\Upsilon(r, s, A) := \prod_{j \in co(s,A)} (z_r - z_{m+j})^{a(r,j)} \prod_{j \in co(r,A)} (z_s - z_{m+j})^{a(s,j)}.$$

3. Let

$$h(A) := \sum_{\theta \in H(A)} \frac{1}{\theta(\delta(z, E(A)))}.$$

**Lemma 5** *Let the notation be as above. Assume that $A$ satisfies the admissibility requirements (1) and (2). Then the $t$-order of $\psi(h(A))$ is $-2a|co(A)|$.*

*Proof* Let $\mu(z) := \delta(z, E(A))$ and let $\varepsilon(i, j)$ denote the $(i, j)$th entry of $E(A)$. Then

$$\mu(z) := \prod_{(i,j) \in \pi[N]} (z_i - z_j)^{\varepsilon(i,j)}.$$

Clearly, $\mu(z) \in \mathbb{Q}[z]$ and

$$\mu(z) = \left( \prod_{r=1}^{m} \alpha(r, A)^{2a} \beta(r, A) \right) \prod_{1 \leq r < s \leq m} \Upsilon(r, s, A).$$

Define $g \in \mathbb{Q}[x]$ by setting

$$g(x_1, \ldots, x_N) := \prod_{r=1}^{m} \prod_{j \in co(r,A)} (x_r - x_{m+j}).$$

For $\theta \in S_N$, define $\theta(g(x))$ to be the polynomial $g(x_{\theta(1)}, \ldots, x_{\theta(N)})$.
    For $\theta \in G(A)$, let $Q_\theta(t, y, T, x) := \psi(\theta(\mu(z)))$ and let

$$P_\theta(t, y, T, x) := \prod_{r=1}^{m} \psi(\theta(\beta(r, A))) \prod_{1 \leq r < s \leq m} \psi(\theta(\Upsilon(r, s, A))).$$

Then, we have

$$Q_\theta(t, y, T, x) = P_\theta(t, y, T, x) \prod_{r=1}^{m} \psi(\theta(\alpha(r, A)))^{2a}.$$

As before, let *id* denote the identity permutation. Then, set $b := |co(A)|$ and observe that

$$Q_{id}(t, y, T, x) = t^{2ab} g(x)^{2a} P_{id}(t, y, T, x)$$

and $P_{id}(0, 0, T, x) = c \cdot pol(A, T)$, where $c \in \{-1, 1\}$. Moreover, if $\theta \in H(A)$ and $\tau \in grp(A)$ denotes the permutation induced by $\theta$, then

$$Q_\theta(t, y, T, x) = t^{2ab} \theta(g(x))^{2a} P_\theta(t, y, T, x)$$

and $P_\theta(0, 0, T, x) := c \cdot \tau(pol(A, T))$. Most importantly, note that $c$ is nonzero and independent of the choice of $\theta$. It is straightforward to verify that

$$t^{2ab}\psi(h(A)) := \sum_{\theta \in H(A)} \frac{t^{2ab}}{Q_\theta(t, y, T, x)} = \sum_{\theta \in H(A)} \frac{1}{\theta(g(x))^{2a} P_\theta(t, y, T, x)}.$$

Each summand of the sum on the extreme right is well defined at $t = y = 0$. Now for $f \in rat(A, T)$, let

$$H_f(A) := \left\{ \theta \in H(A) \,\middle|\, f = \frac{1}{\theta^*(pol(A, T))} \right\},$$

where $\theta^* \in grp(A)$ denotes the permutation induced by $\theta$. Evaluation of $t^{2ab}\psi(h(A))$ at $t = y = 0$ yields

$$(*) \qquad c \cdot \sum_{f \in rat(A,T)} \left( \sum_{\theta \in H_f(A)} \frac{f}{\theta(g(x))^{2a}} \right).$$

By (i) of Theorem 7, the inner sum is nonzero provided it is a nonempty sum. In particular, this inner sum is nonzero when $f = 1/pol(A, T)$ (since in that case $id$ is in $H_f(A)$). By hypothesis, $A$ satisfies the admissibility condition (2) and thus $rat(A, T)$ is a $\mathbb{Q}$-linearly independent subset of $\mathbb{Q}(T)$. Hence, $rat(A, T)$ is also a $\mathbb{Q}(x)$-linearly independent subset of $\mathbb{Q}(T, x)$. So, the $t$-order of $t^{2ab}\psi(h(A))$ is 0. It follows that $\psi(h(A))$ has $t$-order $-2ab = -2a|co(A)|$. $\qquad\square$

**Definitions** We continue to use the notation introduced above. Recall that $A := [a_{ij}]$ is an $m \times (N - m)$ matrix with nonnegative integer entries $a(i, j)$.

1. For $\theta \in S_N$ and $1 \le r \le m$, define

$$R_r(\theta, A) := \left\{ i \,\middle|\, 1 \le i \le m \text{ and } \theta(i) \in C(r, A) \right\},$$
$$K_r(\theta, A)) := \left\{ i \,\middle|\, m + 1 \le i \le N \text{ and } \theta(i) \in C(r, A) \right\}.$$

2. Let

$$h^*(A) := \sum_{\theta \in G(A) \setminus H(A)} \frac{1}{\theta(\delta(z, E(A)))}.$$

**Lemma 6** *Let the notation be as above. Assume that $A$ satisfies the admissibility requirements (1) and (3). Then the $t$-order of $\psi(h^*(A))$ is strictly greater than $-2a|co(A)|$.*

*Proof* Let $\mu(z) := \delta(z, E(A))$ and let $\varepsilon(i, j)$ denote the $(i, j)$th entry of $E(A)$. Then, for $(i, j) \in \pi[N]$, we have

$$\varepsilon(i,j) = \begin{cases} a(i, j - m) & \text{if } 1 \leq i \leq m \text{ and } m + 1 \leq j \leq N, \\ \\ 0 & \text{otherwise.} \end{cases}$$

Now fix a permutation $\theta \in G(A)$ such that $\theta$ is not in $H(A)$. Observe that for $1 \leq r \leq m$, the sets $R_r(\theta, A)$ and $K_r(\theta, A)$ partition $\theta^{-1}(C(r, A))$. In particular,

$$|R_r(\theta, A)| + |K_r(\theta, A)| = 1 + |co(r, A)|.$$

Also, note that

$$\theta(\mu(z)) = \pm \prod_{(i,j) \in \pi[N]} (z_i - z_j)^{\varepsilon(\theta^{-1}(i),\, \theta^{-1}(j))}.$$

For $(i, j) \in \pi[N]$, it is straightforward to verify that the $t$-order of $\psi(z_i - z_j)$ is positive if and only if $(i, j)$ is in $\pi(C(r, A))$ for some $1 \leq r \leq m$. Hence, the $t$-order of $\psi(\theta(\mu(z)))$ is

$$\sum_{r=1}^{m} \sum_{(i,j) \in \pi(C(r,A))} \varepsilon(\theta^{-1}(i),\, \theta^{-1}(j)) = \sum_{r=1}^{m} \sum_{(i,j) \in \pi(\theta^{-1}(C(r,A)))} \varepsilon(i, j).$$

Moreover, by focusing on the nonzero summands of the sum on the extreme right above, we infer that

$$\sum_{(i,j) \in \pi(\theta^{-1}(C(r,A)))} \varepsilon(i, j) = \sum_{(i,j) \in R_r(\theta,A) \times K_r(\theta,A)} a(i, j - m).$$

For $1 \leq r \leq m$, letting $M_r(\theta)$ denote the (possibly empty) submatrix of $A$ determined by the row-set $R_r(\theta, A)$ and the column-set $K_r(\theta, A)$, we clearly have

$$\|M_r(\theta)\| = \sum_{(i,j) \in R_r(\theta,A) \times K_r(\theta,A)} a(i, j - m).$$

From the above observations, it is evident that

$$\text{the } t\text{-order of } \psi(\theta(\mu(z))) = \|M_1(\theta)\| + \cdots + \|M_m(\theta)\|.$$

If for some $r$, the corresponding $M_r(\theta)$ is empty, then $\|M_r(\theta)\| = 0$. If for some $r$, the corresponding $M_r(\theta)$ has only one row, then $\|M_r(\theta)\| \leq 2a|co(r, A)|$. If for some $r$, the corresponding $M_r(\theta)$ has a single column and at least two rows, then the admissibility-condition (1) implies that $\|M_r(\theta)\| < 2a|co(r, A)|$. If for some $r$, the corresponding $M_r(\theta)$ has two or more rows as well as two or more columns, then the admissibility-condition (3) guarantees that $\|M_r(\theta)\| < 2a|co(r, A)|$. Consequently,

$$(**) \qquad \text{the } t\text{-order of } \psi(\theta(\mu(z))) \; = \; \sum_{r=1}^{m} \|M_r(\theta)\| \; \leq \; 2a|co(r, A)|.$$

Furthermore, in view of the admissibility-condition (1) satisfied by $A$, the $\leq$ of $(**)$ is actually $=$ only when $M_r(\theta)$ is a row-matrix with each entry equal to $2a$ for each $r$ with $1 \leq r \leq m$, i.e., only when there is a permutation $\tau$ of $\{1, \ldots, m\}$ such that for $1 \leq r \leq m$,

$$R_r(\theta, A) \; = \; \{\tau(r)\} \quad \text{and} \quad \theta^{-1}(C(r, A)) \subseteq C(\tau(r), A).$$

Since (by our choice) $\theta$ is not in $H(A)$ and $H(A)$ is a subgroup of $S_N$, the permutation $\theta^{-1}$ is not in $H(A)$. Hence $(**)$ must be a strict inequality. Thus we have proved that for each $\theta \in G(A) \setminus H(A)$, the polynomial $\psi(\theta(\mu(z)))$ has $t$-order strictly less than $2a|co(A)|$, i.e., the rational function $1/\psi(\theta(\mu(z)))$ has $t$-order strictly greater than $-2a|co(A)|$. But then, the $t$-order of the rational function $\psi(h^*(A))$ is also strictly greater than $-2a|co(A)|$. $\qquad\square$

Thanks to the above definitions and lemmas, we can now state and prove an important nonvanishing theorem (see Theorem 3 of [12]).

**Theorem 10** *As before, $k$ is a field containing $\mathbb{Q}$, $z_1, \ldots, z_N$ are indeterminates and $z$ stands for $(z_1, \ldots, z_N)$. Let $m$, $N$ be integers such that $1 \leq m \leq N - 2$. Let $A := [a(i, j)]$ be an $m \times (N - m)$ matrix with nonnegative integer entries $a(i, j)$. Assume that*

*(i) $\max(A) = 2a$, where $a$ is a positive integer, and*
*(ii) $A$ is admissible.*

*Then, $Symm_N \left( \frac{1}{\delta(z, E(A))} \right) \neq 0$.*

*Proof* Let $\mu(z) := \delta(z, E(A))$. We continue to use the notation introduced above. First, note that

$$\psi \left( \sum_{\theta \in G(A)} \frac{1}{\theta(\mu(z))} \right) \; = \; \psi(h(A)) + \psi(h^*(A)).$$

By Lemma 5, the $t$-order of $\psi(h(A))$ is the strictly negative integer $-2a|co(A)|$ whereas, by Lemma 6, the $t$-order of $\psi(h^*(A))$ is strictly greater than $-2a|co(A)|$. Hence, $\psi(h(A)) + \psi(h^*(A)) \neq 0$. In particular, $h(A) + h^*(A) \neq 0$. If $G(A) = S_N$, then

$$Symm_N \left( \frac{1}{\mu(z)} \right) \; = \; h(A) + h^*(A) \neq 0,$$

as asserted. Henceforth, assume that $G(A) \neq S_N$. Obviously,

$$Symm_N \left( \frac{1}{\mu(z)} \right) = h(A) + h^*(A) + \sum_{\theta \in S_N \setminus G(A)} \frac{1}{\theta(\mu(z))}.$$

By (iv) of Lemma 4, the total $z$-degree of $\psi(\theta(\mu(z)))$ is $\geq 1$ if and only if $\theta \in S_N \setminus G(A)$. Hence, by the elementary properties of the $z$-degree, the rational function

$$\sum_{\theta \in S_N \setminus G(A)} \frac{1}{\psi(\theta(\mu(z)))}$$

has total $z$-degree at most $-1$. Since $\psi(h(A)) + \psi(h^*(A))$ is a nonzero member of the field $k(t, y, T, x)$, its total $z$-degree is 0. Consequently,

$$\psi \left( Symm_N \left( \frac{1}{\mu(z)} \right) \right) = \psi(h(A)) + \psi(h^*(A)) + \sum_{\theta \in S_N \setminus G(A)} \frac{1}{\psi(\theta(\mu(z)))}$$

has total $z$-degree 0. In particular, $Symm_N \left( \frac{1}{\mu(z)} \right) \neq 0$.                     $\square$

**Corollary** *Let $m$, $N$ be integers such that $1 \leq m \leq N - 2$ and let $A := [a(i, j)]$ be an $m \times (N - m)$ matrix with nonnegative integer entries $a(i, j)$.*

1. *Assume that the following holds.*

   (i) *There is a positive integer $a$ such that $max(A) = 2a$ and $co(r, A) \neq \emptyset$ for $1 \leq r \leq m$.*

   (ii) *For $1 \leq j \leq N - m$, we have*

   $$\sum_{i=1}^{m} a(i, j) \leq 2a,$$

   *i.e., each column-sum of $A$ is at most $2a$.*

   *Then, letting $E(A) \in E(N)$ be defined as in the above theorem, we have*

   $$Symm_N \left( \frac{1}{\delta(z, E(A))} \right) \neq 0.$$

2. *Assume that the following holds.*

   (i) *There is a positive integer $a$ such that $max(A) = 2a$.*

   (ii) *$|co(r, A)| = 1$ for $1 \leq r \leq m$ and*

   $$co(r, A) \cap co(s, A) = \emptyset \quad for \ 1 \leq r < s \leq m.$$

   (iii) *There is a nonnegative integer $b < 2a$ such that for $1 \leq i, r \leq m$ with $i \neq r$ and $j \in co(r, A)$, we have $a(i, j) = b$.*

*Then, letting $E(A) \in E(N)$ be defined as in the above theorem, we have*

$$Symm_N \left( \frac{1}{\delta(z, E(A))} \right) \neq 0.$$

3. *Let $a, b, c, r, s$ be positive integers such that $b < 2a \leq 2c$ and $r \leq s \leq N - 1$. Suppose $A := [u_1, \ldots, u_{N-1}]$ is the $1 \times N - 1$ matrix such that $u_i := 2a$ for $1 \leq i \leq r$, $u_i := b$ for $r + 1 \leq i \leq s$, and $u_i = 0$ for $s + 1 \leq i \leq N - 1$. Let $E(A) \in E(N)$ be defined as in the above theorem. Then, letting $E_{(r,s)}(N; a, b, c) := 2cD_N - E(A)$, we have $Symm_N(\delta(z, E_{(r,s)}(N; a, b, c))) \neq 0$.*

*Proof* To prove the first two assertions, it suffices to show that under their respective hypotheses, $A$ is admissible. Firstly, assume that $A$ satisfies the requirements of assertion 1. It follows from the hypothesis (ii) of assertion 1 that given $j \in co(A)$, there is only one nonzero entry in the $j$th column of $A$ and that nonzero entry is $2a$. So, we have $\nu(r, s, A) = 0$ for $1 \leq r < s \leq m$ and hence

$$rat(A, T) = \left\{ T_{\tau(1)}^{-b(1,A)} \cdot \cdots \cdot T_{\tau(m)}^{-b(m,A)} \;\middle|\; \tau \in grp(A) \right\}.$$

Clearly, $rat(A, T)$ is a $\mathbb{Q}$-linearly independent subset of $\mathbb{Q}(T)$. Suppose $M$ is a $p \times q$ submatrix of $A$, where $p \geq 2$. By hypothesis (iii) of assertion 1, we have $\|M\| \leq 2aq$ and $2aq < 2a(q + 1) \leq 2a(p + q - 1)$. Thus, $A$ is readily seen to be an admissible matrix.

Secondly, suppose $A$ satisfies the requirements of assertion 2. By hypothesis (iii) of assertion 2, $\nu(r, s, A) = 2b$ for $1 \leq r < s \leq m$. Consequently,

$$rat(A, T) = \left\{ \Delta(T)^{-b} \cdot T_{\tau(1)}^{-b(1,A)} \cdot \cdots \cdot T_{\tau(m)}^{-b(m,A)} \;\middle|\; \tau \in grp(A) \right\}.$$

Now it is straightforward to verify that $rat(A, T)$ is a $\mathbb{Q}$-linearly independent subset of $\mathbb{Q}(T)$. Consequently, $A$ is admissible.

Lastly, consider assertion 3. Since $\delta(z, 2cD_n) = \Delta(z)^c$ is symmetric in the variables $z_1, \ldots, z_N$, we have

$$Symm_N(\delta(z, 2cD_n - E(A))) = \Delta(z)^c \cdot Symm_N \left( \frac{1}{\delta(z, E(A))} \right).$$

Observe that $A$ satisfies the hypotheses of assertion 1 and hence

$$Symm_N \left( \frac{1}{\delta(z, E(A))} \right) \neq 0.$$

So, the product on the right of the above equality is indeed nonzero.                  □

**Definitions** Let $m, N$ be integers such that $m \geq 2$ and $N \geq m + 2$. Let $t, T_1, \ldots, T_m$ and $x_1, \ldots, x_N$ be indeterminates, let $x$ stand for $(x_1, \ldots, x_N)$ and let $T$ stand for $(T_1, \ldots, T_m)$.

1. Define
$$\pi[m, N] := \{1, \ldots, m\} \times \{m + 1, \ldots, N\},$$
$$B_m := \{(r, m + r) \mid 1 \leq r \leq m\}.$$

2. For $\theta \in S_N$, let
$$B_m(N, \theta) := \{(i, j) \in \pi[N] \mid \theta(i, j) \in B\}.$$

3. Let
$$G(m, N) := \{\theta \in S_N \mid B_m(N, \theta) = B_m\}.$$

4. Let $\sigma : k[z] \to k[t, T, x]$ be the $k$-homomorphism of rings defined by

$$\sigma(z_i) := \begin{cases} tx_i + T_i & \text{if } 1 \leq i \leq m, \\ \\ tx_i + T_{i-m} & \text{if } m + 1 \leq i \leq N. \end{cases}$$

5. For $(i, j) \in \{1, \ldots, m\} \times \{1, \ldots, m\}$, let

$$q_1(i, j) := (z_i - z_{m+j})(z_j - z_{m+i}),$$
$$q_2(i, j) := (z_i - z_j)(z_{m+j} - z_{m+i}).$$

*Remark* The above defined map $\sigma$ is easily seen to be an injective homomorphism of rings and hence it naturally extends to an injective field-homomorphism $k(z) \to k(t, T, x)$. To economize notation, this field-homomorphism is also denoted by $\sigma$. The definitions and notation introduced above allow us to prove yet another result ensuring nontriviality of certain symmetrizations (see the fourth corollary to Theorem 3 of [12]).

**Theorem 11** *Let $m, N$ be integers such that $m \geq 2$ and $N = 2m$. Assume $k$ is a field containing $\mathbb{Q}$, $z_1, \ldots, z_N$ are indeterminates and let $z$ stand for $(z_1, \ldots, z_N)$. Let $A := [a(i, j)]$ be an $m \times m$ matrix with nonnegative integer entries $a(i, j)$ satisfying the following two requirements.*

*(i) $a(i, j) = a(j, i)$ for $1 \leq i < j \leq m$, i.e., $A$ is symmetric.*
*(ii) There are positive integers $a, a_1, \ldots, a_m$ such that*

$$a(i, i) = 2a_i \geq 2a \quad \text{for } 1 \leq i \leq m \text{ and } \quad 2a > a(i, j) \quad \text{for } 1 \leq i < j \leq m.$$

*Let $E(A) \in E(N)$ be the matrix given in block-form by*

$$E(A) := \begin{bmatrix} 0 & A \\ A^T & 0 \end{bmatrix}.$$

*Then, $Symm_N(\delta(z, -E(A))) \neq 0$.*

*Proof* Define $\mu(z) := \delta(z, E(A))$. In view of our hypothesis (i),

$$\mu(z) := \prod_{1 \leq i \leq m} (z_i - z_{m+i})^{2a_i} \prod_{1 \leq i < j \leq m} [(z_i - z_{m+j})(z_j - z_{m+i})]^{a(i,j)}.$$

Note that $B_m \subseteq \pi[m, N]$ and $B_m(N, id) = B_m$, where $id$ is the identity permutation. Clearly, the identity permutation is an element of $G(m, N)$. For $(i, j) \in \{1, \ldots, m\} \times \{1, \ldots, m\}$, we have $q_1(i, j) = q_1(j, i), q_2(i, j) = q_2(j, i)$, and $q_2(i, i) = 0$. Evidently,

$$\mu(z) = \prod_{1 \leq i \leq m} q_1(i, i)^{a_i} \prod_{1 \leq i < j \leq m} q_1(i, j)^{a(i,j)}.$$

Fix $\theta \in G(m, N)$ and $(i, j) \in \pi[m]$. Clearly, $\theta(p, m + p) \in B_m$ for all $(p, m + p) \in B_m$ and hence $\{\theta(i), \theta(m + i)\} = \{r, m + r\}$ for some $1 \leq r \leq m$. Likewise, $\{\theta(j), \theta(m + j)\} = \{s, m + s\}$ for some $1 \leq s \leq m$. Since $i \neq j$, we have $\{r, m + r\} \cap \{s, m + s\} = \emptyset$. Observe that $\sigma(\theta(q_1(i, i))) = t^2(x_i - x_{m+i})^2$ and $\theta(q_1(i, j)) \in \{q_1(r, s), q_2(r, s)\}$. Thus, if $i \neq j$, then for $1 \leq p \leq 2$, the polynomial

$$\sigma(\theta(q_p(i, j))) - (T_r - T_s)^2 \in k[t, T, x]$$

is divisible by $t$.

For $\theta \in S_N \setminus G(m, N)$, we have

$$|B_m(N, \theta) \cap B_m| \leq (m - 1) \quad \text{and} \quad |B_m(N, \theta) \cap \pi[m, N]| \leq m.$$

Using our hypothesis (ii), we deduce that the $t$-order of $\sigma(\theta(\mu(z)))$ is strictly less than $d := 2(a_1 + \cdots + a_m)$. On the other hand, for $\theta \in G(m, N)$, there are polynomials $P_\theta(x) \in k[x]$ and $Q_\theta(t, x, T) \in k[t, x, T]$ such that

$$\sigma(\theta(\mu(z))) = t^d P_\theta(x)^2 Q_\theta(t, x, T).$$

Furthermore, from what was observed above, there is a polynomial $h_\theta(T) \in k[T]$ such that $Q_\theta(0, x, T) = h_\theta(T)^2$. Let

$$v(t, x, T) := \sum_{\theta \in S_N \setminus G(m, N)} \frac{1}{\sigma(\theta(\mu(z)))} \quad \text{and} \quad w(t, x, T) := \sum_{\theta \in G(m, N)} \frac{t^d}{\sigma(\theta(\mu(z)))}.$$

Then $\sigma(Symm_N(\delta(z, -E(A)))) = v(t, x, T) + t^{-d} w(t, x, T)$.

Now first note that the $t$-order of $v(t, x, T)$ is strictly greater than $-d$. Secondly, since

$$w(0, x, T) = \sum_{\theta \in G} \left( \frac{1}{P_\theta(x) h_\theta(T)} \right)^2,$$

we have $w(0, x, T) \neq 0$ by assertion (i) of Theorem 7. In particular, $w(t, x, T) \neq 0$. It follows that the $t$-order of $v(t, x, T) + t^{-d} w(t, x, T)$ is exactly $-d$ and hence $Symm_N(\delta(z, -E(A))) \neq 0$ as asserted.  $\square$

*Example* We present an example which shows that although the above theorem is similar in spirit to Theorem 10, it does offer something essentially different. For example, consider the $6 \times 6$ symmetric matrix

$$A := \begin{bmatrix} 2 & 0 & 1 & 1 & 1 & 1 \\ 0 & 2 & 1 & 1 & 1 & 1 \\ 1 & 1 & 2 & 0 & 1 & 1 \\ 1 & 1 & 0 & 2 & 1 & 1 \\ 1 & 1 & 1 & 1 & 2 & 1 \\ 1 & 1 & 1 & 1 & 1 & 2 \end{bmatrix}.$$

Clearly, $max(A) = 2$ and $co(r, A) = \{r\}$ for $1 \leq r \leq 6$. So, $grp(A) = S_6$ and $b(r, A) = 0$ for $1 \leq r \leq 6$. Note that $A$ satisfies requirements (1) and (3) for admissibility. It is straightforward to verify that

$$\nu(r, s, A) = \begin{cases} 0 & \text{if either } 1 = r < s = 2 \text{ or } 3 = r < s = 4, \\ 2 & \text{otherwise.} \end{cases}$$

As a consequence, we have

$$pol(A, T) = (T_1 - T_2)^{-2} \cdot (T_3 - T_4)^{-2} \cdot \Delta, \quad \text{where} \quad \Delta := \prod_{1 \leq r < s \leq 6} (T_r - T_s)^2.$$

Since $\Delta$ is symmetric in $T_1, \ldots, T_6$, the $\mathbb{Q}$-linear independence of $rat(A, T)$ is equivalent to the $\mathbb{Q}$-linear independence of the set

$$R := \left\{ (T_{\theta(1)} - T_{\theta(2)})^2 (T_{\theta(3)} - T_{\theta(4)})^2 \,\middle|\, \theta \in S_6 \right\}.$$

Letting $f_1$ denote the polynomial

$$(T_1 - T_2)^2 (T_3 - T_4)^2 + (T_2 - T_3)^2 (T_1 - T_6)^2 + (T_2 - T_5)^2 (T_1 - T_4)^2 +$$
$$(T_5 - T_6)^2 (T_1 - T_2)^2 + (T_3 - T_6)^2 (T_1 - T_4)^2 + (T_4 - T_5)^2 (T_1 - T_6)^2 +$$
$$(T_4 - T_5)^2 (T_2 - T_3)^2 + (T_3 - T_6)^2 (T_2 - T_5)^2 + (T_5 - T_6)^2 (T_3 - T_4)^2$$

and letting $f_2$ denote the polynomial

$$(T_2 - T_3)^2 (T_1 - T_4)^2 + (T_3 - T_6)^2 (T_1 - T_2)^2 + (T_4 - T_5)^2 (T_1 - T_2)^2 +$$
$$(T_2 - T_5)^2 (T_1 - T_6)^2 + (T_3 - T_4)^2 (T_1 - T_6)^2 + (T_5 - T_6)^2 (T_1 - T_4)^2 +$$
$$(T_3 - T_4)^2 (T_2 - T_5)^2 + (T_5 - T_6)^2 (T_2 - T_3)^2 + (T_3 - T_6)^2 (T_4 - T_5)^2 ,$$

it is clear that each of $f_1, f_2$ is a $\mathbb{Z}$-linear combination of (pairwise distinct) elements of the set $R$. Now it is straightforward to verify that $f_1 = f_2$ and hence $R$ is $\mathbb{Q}$-linearly dependent. Thus, $A$ is not admissible. On the other hand, taking $a = a_1 = a_2 = a_3 = a_4 = a_5 = a_6 = 1$, in (ii), we deduce that $A$ satisfies the hypotheses of Theorem 11.

**Definitions** Let $r, s$ be positive integers.

1. Let $\mathbb{M}(r, s)$ be the set of $r \times s$ matrices with nonnegative integer entries.
2. Let $\mathbb{M}_+(r, s)$ be the set of $r \times s$ matrices with positive integer entries.
3. Let $\mathbb{M}_2(r, s)$ be the set of $r \times s$ matrices whose entries are nonnegative even integers.
4. Given nonnegative integers $d$ and $\lambda$, define $\mathbb{M}(r, s, d, \lambda)$ to be the subset of all $A := [a_{ij}] \in \mathbb{M}(r, s)$ such that $\|A\| = \lambda$,

$$\sum_{j=1}^{s} a_{ij} \leq d \quad \text{for } 1 \leq i \leq r \text{ and} \quad \sum_{i=1}^{r} a_{ij} \leq d \quad \text{for } 1 \leq j \leq s.$$

5. Define
$$\mathbb{M}_+(r, s, d, \lambda) := \mathbb{M}(r, s, d, \lambda) \cap \mathbb{M}_+(r, s) \quad \text{and}$$
$$\mathbb{M}_2(r, s, d, \lambda) := \mathbb{M}(r, s, d, \lambda) \cap \mathbb{M}_2(r, s).$$

*Remark* Given $A \in \mathbb{M}(r, s)$, it is clear that $A$ is in $\mathbb{M}(r, s, d, \lambda)$ (respectively, in $\mathbb{M}_+(r, s, d, \lambda)$, in $\mathbb{M}_2(r, s, d, \lambda)$) if and only if $A^T$ is in $\mathbb{M}(s, r, d, \lambda)$ (respectively, in $\mathbb{M}_+(s, r, d, \lambda)$, in $\mathbb{M}_2(s, r, d, \lambda)$).

**Lemma 7** *Suppose $d, \lambda$ are nonnegative integers and $r, s$ are positive integers.*

(i) $\mathbb{M}(r, s, d, \lambda)$ *is nonempty if and only if $\lambda \leq min\{rd, sd\}$.*
(ii) $\mathbb{M}_+(r, s, d, \lambda)$ *is nonempty if and only if $d \geq max\{r, s\}$ and*

$$rs \leq \lambda \leq min\{rd, sd\}.$$

(iii) $\mathbb{M}_2(r, s, d, \lambda)$ *is nonempty if and only if $\lambda$ is even and*

$$\frac{\lambda}{2} \leq min\left\{r\left\lfloor \frac{d}{2} \right\rfloor, s\left\lfloor \frac{d}{2} \right\rfloor\right\}.$$

(iv) *Suppose $r \geq 2$ and $2\lambda \leq rd$. Then there exists an $A \in E(r, \leq d)$ such that $\|A\| = 2\lambda$.*

*Proof* If $A \in \mathbb{M}(r, s, d, \lambda)$, then obviously $\|A\| \leq min\{rd, sd\}$. Conversely, assume that $\lambda$ is a nonnegative integer such that $\lambda \leq min\{rd, sd\}$. We prove the existence of a matrix $A \in \mathbb{M}(r, s, d, \lambda)$ by induction on $\max\{r, s\}$. If $r = s = 1$, then our assertion clearly holds. Henceforth, suppose $\max\{r, s\} \geq 2$. Note that we may assume $r \leq s$, without loss of generality. If $\lambda \leq \min\{rd, (s - 1)d\}$, then our induction hypothesis ensures the existence of $B \in \mathbb{M}(r, s - 1, d, \lambda)$; choosing such a $B$ and by letting $A$ be the block-matrix $[B, 0]$, we have $A \in \mathbb{M}(r, s, d, \lambda)$. If $\lambda > \min\{rd, (s - 1)d\}$, then we must have $r = s$ and $(r - 1)d < \lambda \leq rd$. Hence $0 < (\lambda - (r - 1)d) \leq d$. Since $r \geq 2$ in this case, our induction hypothesis ensures the existence of $B \in \mathbb{M}(r - 1, r - 1, d, (r - 1)d)$. Choose such a $B$ and let $A$ be the block-matrix with rows $[B, 0]$ and $[0, \lambda - (r - 1)d]$. Then, it is easy to verify that $A$ is in $\mathbb{M}(r, r, d, \lambda)$. This establishes (i).

If $A \in \mathbb{M}_+(r, s, d, \lambda)$, then $d \geq \max\{r, s\}$ and $rs \leq \|A\| \leq \min\{rd, sd\}$. Conversely, assume that $d \geq \max\{r, s\}$ and $\lambda$ is a positive integer such that $rs \leq \lambda \leq \min\{rd, sd\}$. We prove the existence of a matrix $A \in \mathbb{M}_+(r, s, d, \lambda)$ by induction on $\max\{r, s\}$. If $r = s = 1$, then our assertion clearly holds. Henceforth, suppose $\max\{r, s\} \geq 2$. Again, without loss of generality, we assume that $r \leq s$. First, consider the case where $r < s$. In this case, we have $d \geq s \geq 2$, $r \leq (s - 1)$, and $rs \leq \lambda \leq rd$. Observe that $r(s - 1) \leq (\lambda - r) \leq r(d - 1)$ and hence our induction hypothesis ensures the existence of $B \in \mathbb{M}_+(r, s - 1, d - 1, \lambda - r)$. Pick such a $B$ and let $A$ be the block-matrix $[B, C]$, where $C$ denotes the $r \times 1$ column with all entries 1. Then $A$ is easily seen to be in $\mathbb{M}_+(r, s, d, \lambda)$. Next, consider the case where $r = s \geq 2$. If $\lambda \leq (r - 1)(d - 1) + 2r - 1$, then since $d - 1 \geq r - 1$, $0 < (r - 1)^2 \leq (\lambda - 2r + 1) \leq (r - 1)(d - 1)$, our induction hypothesis ensures the existence of $B \in \mathbb{M}_+(r - 1, r - 1, d - 1, \lambda - 2r + 1)$. Pick such a $B$ and let $A$ be the block-matrix with rows $[B, C]$ and $[C^T, 1]$, where $C$ denotes the $(r - 1) \times 1$ column with all entries 1. It follows that $A \in \mathbb{M}_+(r, r, d, \lambda)$. Lastly, if $r = s \geq 2$ and $2r + (r - 1)(d - 1) \leq \lambda \leq rd$, then $2 \leq (\lambda - rd - r + d + 1) \leq d - (r - 1)$. Now, using our induction hypothesis, pick a $B \in \mathbb{M}_+(r - 1, r - 1, d - 1, (r - 1)(d - 1))$ and let $A$ be the block-matrix with rows $[B, C]$ and $[C^T, \lambda - rd - r + d + 1]$, where $C$ denotes the $(r - 1) \times 1$ column with all entries 1. As before, one easily verifies that $A$ is a member of $\mathbb{M}_+(r, r, d, \lambda)$. This proves (ii).

Thirdly, suppose $A := [a_{ij}] \in \mathbb{M}_2(r, s, d, \lambda)$. Then clearly $\lambda$ is an even integer and $(1/2)A \in \mathbb{M}(r, s, \lfloor d/2 \rfloor, \lambda/2)$ and hence by (i), $(\lambda/2) \leq \min\{r, s\} \cdot \lfloor d/2 \rfloor$. Conversely, if $\lambda$ is an even integer satisfying the necessary inequality, then by (i), $\mathbb{M}(r, s, \lfloor d/2 \rfloor, \lambda/2)$ is nonempty. Now for any $B \in \mathbb{M}(r, s, \lfloor d/2 \rfloor, \lambda/2)$, we clearly have $2B \in \mathbb{M}_2(r, s, d, \lambda)$. Thus (iii) holds.

Finally, suppose $r \geq 2$ and $2\lambda \leq rd$. If $d = 0$, then let $A = 0$. Henceforth, assume $d$ to be a positive integer. Firstly, assume $r = 2m$ for a positive integer $m$. Using the fact that $\lambda \leq md$, choose nonnegative integers $a_1, \ldots, a_m$ such that $a_i \leq d$ for $1 \leq i \leq m$ and $a_1 + \cdots + a_m = \lambda$. Letting $A$ be the block-diagonal matrix with $m$ diagonal blocks $a_1 D_2, \ldots, a_m D_2$, it follows that $A \in E(r, \leq d)$ and $\|A\| = 2\lambda$. Next, suppose $r = 2m + 1$ and $d = 2n$, where $m$ and $n$ are positive integers. Using the fact that $\lambda \leq rn$, choose nonnegative integers $a_1, \ldots, a_r$ such that $a_i \leq n$ for $1 \leq i \leq r$ and $a_1 + \cdots + a_r = \lambda$. Let $A$ be the unique $r \times r$ symmetric matrix $[u_{ij}]$, where for $1 \leq i \leq j \leq r$,

$$u_{ij} := \begin{cases} a_i & \text{if } j = i+1, \\ a_r & \text{if } i = 1 \text{ and } j = r, \\ 0 & \text{otherwise.} \end{cases}$$

Then $A$ is easily seen to be in $E(r, \le d)$ and $\|A\| = 2\lambda$. Now suppose $r = 2m+1$ and $d = 2n+1$, where $m$ and $n$ are positive integers. In this case, we have $r \ge 3$ and $\lambda \le rn + m$. If $\lambda \le rn$, then in view of the above argument, we can choose a matrix $A \in E(r, \le (d-1))$ such that $\|A\| = 2\lambda$; observe that $A$ is also in $E(r, \le d)$. So it suffices to restrict to the case where $rn + 1 \le \lambda \le rn + m$. Now, using the fact $0 \le (\lambda - m) \le rn$, choose nonnegative integers $a_1, \ldots, a_r$ such that $a_i \le n$ for $1 \le i \le r$ and $a_1 + \cdots + a_r = \lambda - m$. Let $A$ be the unique $r \times r$ symmetric matrix $[u_{ij}]$, where for $1 \le i \le j \le r$,

$$u_{ij} := \begin{cases} a_i + 1 & \text{if } i \text{ is odd and } j = i+1, \\ a_i & \text{if } i \text{ is even and } j = i+1, \\ a_r & \text{if } i = 1 \text{ and } j = r, \\ 0 & \text{otherwise.} \end{cases}$$

It is straightforward to verify that $A \in E(r, \le d)$ and $\|A\| = 2\lambda$. Thus assertion (iv) is established.                                                                               $\square$.

The following Theorem 12 presents some applications of Theorems 8–9 that allow us to construct various types of semi-invariants in a systematic way.

**Theorem 12** *Let $N$, $d$ and $\lambda$ be positive integers such that $N \ge 3$. As before, let $z_1, \ldots, z_N$ be indeterminates and let $z$ stand for $(z_1, \ldots, z_N)$.*

(i) *Suppose $m$ is a positive integer such that $2(N - d) \le 2m < N$ and $m(N - m) \le \lambda \le md$. Then, there exists a matrix $E \in E(N, \le d)$ such that $Symm_N(\delta(z, E))$ is a nonzero homogeneous polynomial (with integer coefficients) of degree $\lambda$ having $z_i$-degree $\le d$ for $1 \le i \le N$.*

(ii) *Suppose $\lambda$ is even and*

$$\lambda \le 2 \left\lfloor \frac{N}{2} \right\rfloor \left\lfloor \frac{d}{2} \right\rfloor.$$

*Then, there exists a matrix $E \in E(N, \le d)$ such that the entries of $E$ are even numbers and $Symm_N(\delta(z, E))$ is a nonzero homogeneous polynomial (with integer coefficients) of degree $\lambda$ having $z_i$-degree $\le d$ for $1 \le i \le N$.*

(iii) *Suppose $N$, $\lambda$ are even, $2d \ge N$, and*

$$\frac{N^2}{4} \le \lambda \le \frac{Nd}{2}.$$

*Then, there exists a matrix $E \in E(N, \le d)$ such that $Symm_N(\delta(z, E))$ is a nonzero homogeneous polynomial (with integer coefficients) of degree $\lambda$ having $z_i$-degree $\le d$ for $1 \le i \le N$.*

*(iv)  Suppose b, w are nonnegative integers and m is a positive integer such that*

(1) $m < N - m$,
(2) $d \geq (2b + 1)(N - m)$,
(3) $2w \leq min \left\{ Nb, \ m \left\lfloor \frac{d - N + m}{2} \right\rfloor \right\}$   *and*
(4) $\lambda = m(2b + 1)(N - m) + 2w$.

*Then, there exists a matrix $E \in E(N, \leq d)$ such that $Symm_N(\delta(z, E))$ is a nonzero homogeneous polynomial (with integer coefficients) of degree $\lambda$ having $z_i$-degree $\leq d$ for $1 \leq i \leq N$.*

*(v)  Assume there exists an ordered triple $(a, m, u)$ of positive integers such that*

(1) $2 \leq m \leq N - 2$,
(2) $\lambda = mu + am(m - 1)$   *and*
(3) $N - m \leq u \leq min\{(N - m) \lfloor \frac{d}{m} \rfloor, \ d - 2a(m - 1)\}$.

*Then, there exists a matrix $E \in E(N, \leq d)$ such that $Symm_N(\delta(z, E))$ is a nonzero homogeneous polynomial (with integer coefficients) of degree $\lambda$ having $z_i$-degree $\leq d$ for $1 \leq i \leq N$.*

*Proof*  Suppose $m$ is as in (i). Let $m_1 := m$ and $m_2 := N - m$. Then $m_1 < m_2$ and $m_1 + m_2 = N$. Let sets $A_1, A_2$ be defined as in Theorem 8. In view of the hypothesis in (i), assertion (ii) of the above Lemma 7 assures that $\mathbb{M}_+(m, N - m, d, \lambda)$ is nonempty. Choose a matrix

$$C := [c(i, j)] \in \mathbb{M}_+(m, N - m, d, \lambda)$$

and let $E$ be the block-matrix

$$E := \begin{bmatrix} 0 & C \\ C^T & 0 \end{bmatrix}.$$

Then $E$ is in $E(N, \leq d)$. Define $\varepsilon : \pi[N] \to \mathbb{N}$ by setting

$$\varepsilon(i, j) := \begin{cases} 0 & \text{if } (i, j) \in \pi(A_r) \text{ with } 1 \leq r \leq 2, \\ c(i, j - m) & \text{if } (i, j) \in A_1 \times A_2. \end{cases}$$

Then $\varepsilon$ satisfies conditions (1) and (2) of Theorem 8 and letting $\mu(z, \varepsilon)$ be as in Theorem 8, we have $\mu(z, \varepsilon) = \delta(z, E)$. Now (i) follows from (i) of Theorem 8.

Secondly, assume $\lambda$ satisfies the requirements of (ii). Let $m$ be a positive integer such that $2m \leq N$ and $\lambda \leq 2m \lfloor d/2 \rfloor$. Then, assertion (iii) of the above Lemma 7 assures that $\mathbb{M}_2(m, N - m, d, \lambda)$ is nonempty. Choose a matrix

$$C \in \mathbb{M}_2(m, N - m, d, \lambda)$$

and let $E$ be the block-matrix

$$E := \begin{bmatrix} 0 & C \\ C^T & 0 \end{bmatrix}.$$

Then $E$ is in $E(N, \leq d)$ and the entries of $E$ are even numbers. Since $\delta(z, E)$ is a square of a nonzero homogeneous polynomial in $\mathbb{Q}[z_1, \ldots z_N]$, assertion (i) of Theorem 7 allows us to infer that $Symm_N(\delta(z, E)) \neq 0$. Thus (ii) holds.

Next, suppose $N$, $d$, $\lambda$ satisfy the requirements of (iii). Let $m_1 = m_2 = N/2$. Then $m_1 \leq m_2$ and $m_1 + m_2 = N$. Let sets $A_1$, $A_2$ be defined as in Theorem 8. Now assertion (ii) of the above Lemma 7 assures that $\mathbb{M}_+(N/2, N/2, d, \lambda)$ is nonempty. Choose a matrix

$$C := [c(i, j)] \in \mathbb{M}_+(N/2, N/2, d, \lambda)$$

and let $E$ be the block-matrix

$$E := \begin{bmatrix} 0 & C \\ C^T & 0 \end{bmatrix}.$$

Then $E$ is in $E(N, \leq d)$. Define $\varepsilon : \pi[N] \to \mathbb{N}$ by setting

$$\varepsilon(i, j) := \begin{cases} 0 & \text{if } (i, j) \in \pi(A_r) \text{ with } 1 \leq r \leq 2, \\ c\left(i, j - \frac{N}{2}\right) & \text{if } (i, j) \in A_1 \times A_2. \end{cases}$$

Then $\varepsilon$ satisfies conditions (1) and (2) (also (1) and (3)) of Theorem 8 and letting $\mu(z, \varepsilon)$ be as in Theorem 8, we have $\mu(z, \varepsilon) = \delta(z, E)$. Hence (iii) follows from (i) of Theorem 8.

Let $b$, $m$ and $w$ be as in (iv). To begin with, note that $N - m \geq 2$ and $d - m(2b + 1) \geq (2b + 1)$. Define nonnegative integers $d_1$, $d_2$ by

$$d_1 := \min\left\{b, \left\lfloor \frac{d - (2b + 1)(N - m)}{2} \right\rfloor\right\} \quad \text{and} \quad d_2 := b.$$

In view of (3), there are nonnegative integers $w_1$, $w_2$ such that $w = w_1 + w_2$, $2w_1 \leq md_1$, and $2w_2 \leq (N - m)d_2$. Let $C$ be the $m \times (N - m)$ matrix each of whose entries equals $2b + 1$. Let $A_1 \in E(m, \leq d_1)$ be such that $\|A_1\| = 2w_1$. If $m = 1$, then $A_1 = 0$. If $m \geq 2$, then existence of $A_1$ is ensured by (iv) of the above Lemma 7. Let $A_2 \in E(N - m, d_2)$ be such that $\|A_2\| = 2w_2$. Existence of $A_2$ is ensured by (iv) of the above Lemma. Let $E \in E(N, \leq d)$ be the block-matrix

$$E := \begin{bmatrix} 2A_1 & C \\ C^T & 2A_2 \end{bmatrix}.$$

Then, clearly $\|E\| = 2\lambda$. By (i) of Theorem 7, we have

$$Symm_m(\delta(z_1, \ldots, z_m, 2A_1)) \neq 0 \neq Symm_{(N-m)}(\delta(z_1, \ldots, z_{N-m}, 2A_2)).$$

Since each entry of either of $2A_1$, $2A_2$ is at most $2b$ whereas each entry of $C$ is strictly greater than $2b$, Theorem 9 guarantees that $Symm_N(\delta(z, E)) \neq 0$. Thus (iv) stands verified.

Lastly, fix an integer triple $(a, m, u)$ satisfying conditions (1)–(3) listed in (v). Let $C$ be any $m \times (N - m)$ matrix with positive integer entries, having each row-sum exactly $u$ and each column-sum $\leq d$. Using the fact that $u \leq (N - m)\lfloor d/m \rfloor$, choose positive integers $a_1, a_2, \ldots, a_{N-m}$ so that $ma_i \leq d$ for $1 \leq i \leq N - m$ and

$$u = a_1 + a_2 + \cdots + a_{N-m},$$

and let $C$ be the $m \times (N - m)$ matrix each of whose rows equals $[a_1, a_2, \ldots, a_{N-m}]$. Clearly, for any such $C$, we have $\|C\| = mu$. Let $E$ be the block-matrix

$$E := \begin{bmatrix} 2aD_m & C \\ C^T & 0 \end{bmatrix}.$$

It is straightforward to verify that $E$ is in $E(N, \leq d)$ and $\|E\| = 2\lambda$. Now define $q := m + 1, m_i := 1$ for $1 \leq i \leq q - 1$ and $m_q := N - m$. Then $1 \leq m_1 \leq \cdots \leq m_q$ and $m_1 + \cdots + m_q = N$. Let $\varepsilon(i, j)$ be the $(i, j)$th entry of $E$ for $1 \leq i < j \leq N$. Then $\varepsilon$ satisfies the requirements (1) and (2) of Theorem 8 and thus our assertion follows from (i) of Theorem 8. $\qquad\square$

**Corollary**

(i) *Let $\alpha$, $\beta$, $\gamma$ be positive integers such that*

$$(\alpha + 1) \leq \beta \leq 2(\alpha - 1) \quad \text{and} \quad \frac{\gamma(\beta - \alpha)}{\beta - 1} \text{ is an integer.}$$

*Then there exists a nonzero binary invariant $I_{\alpha\beta\gamma}$ of type $(2\alpha, \beta\gamma)$. Moreover, if each of $\alpha$, $\beta$, $\gamma$ is odd, then $I_{\alpha\beta\gamma}$ is a skew binary invariant, i.e., a nonzero binary invariant of odd (total) degree in $z_1, \ldots, z_N$.*

(ii) *Let $s$ and $t$ be positive integers such that $s + 1 \leq t \leq 2s - 1$. Then to each positive integer pair $(r, v)$ corresponds a nonzero binary invariant $I$ of type $(2(2sv + 1), rt(2tv + 1))$; moreover, if $r$ and $t$ both are odd, then $I$ is a skew binary invariant.*

*Proof* Define $N := 2\alpha, d := \beta\gamma, \lambda := \alpha\beta\gamma$,

$$a := \frac{\gamma(\beta - \alpha)}{\beta - 1}, \quad m := \beta, \quad \text{and} \quad u := \gamma(2\alpha - \beta).$$

Then, it is straightforward to verify that our $\lambda$ and $(a, m, u)$ satisfy the hypotheses of assertion (v) of Theorem 11. Let $E$ be defined as in the proof of that assertion, where $C$ has positive integer entries, each row-sum equal to $\gamma(2\alpha - \beta)$ and each column-sum less than or equal to $\beta\gamma$, e.g., we may let $C$ be the $\beta \times (2\alpha - \beta)$ matrix having each entry equal to $\gamma$. Then, $E \in E(N, \leq d)$, $\|E\| = 2\alpha\beta\gamma$ and by (v), we have $Symm_N(\delta(z, E)) \neq 0$. Moreover, since each row-sum of $C^T$ is exactly $d$, we have $E \in E(N, d)$. Thus (i) follows from assertion (i) of Theorem 5. Assertion (ii) follows from (i) by specializing $\alpha$ to $(2sv + 1)$, $\beta$ to $(2tv + 1)$, and $\gamma$ to $rt$.      □

*Remarks*

1. Suppose $f$ and $g$ are rational functions of $z_1, \ldots, z_N$ (with coefficients in an integral domain) such that $Symm_N(f)$ and $Symm_N(g)$ both are nonzero. Then, there exists $\sigma \in S_N$ such that $Symm_N(f \cdot \sigma(g))$ is nonzero. This useful observation can be used in conjunction with our theorems when $N$ is relatively small. In the most general case, there is no known procedure to find the needed permutation $\sigma$.
2. Suppose $f$ is a rational function of $z_1, \ldots, z_m$ and $g$ is a rational function of $z_{m+1}, \ldots, z_N$ (with coefficients in an integral domain) such that $Symm_m(f) = 0$. Then, $Symm_N(fg) = 0$. This observation can be used to show that if $E := M(m, n, (2b + 1)/2, 0)$, then $Symm_N(\delta(z, E)) = 0$.
3. Nonzero-ness of a symmetrization of a rational function in $k(z_1, \ldots, z_N)$ can be established by means of an appropriate numerical substitution; an example of this can be found in the proof of Proposition 1.10 of [13].
4. For an $E \in E(N)$, there is a graph-theoretic necessary and sufficient criterion established in [14] which is equivalent to the nonzero-ness of $Symm_N(\delta(z, E))$. Nevertheless, how this criterion can be put to work remains unknown at present.

Our Theorems 7–12 facilitate a large number of constructions of binary invariants and semi-invariants. The following list of examples provides a small sample of such constructions.

*Examples*

1. Let $N := 6$ and $d := 10$. Consider integers $\lambda$ such that $2 \leq \lambda \leq 30$. With the only exceptions of $\lambda = 3$ and $\lambda = 29$, the above theorems provide constructions of $E_\lambda \in E(6, \leq 10)$, with $2\lambda = \|E_\lambda\|$ and $Symm_6(\delta(z, E_\lambda)) \neq 0$. For $\lambda$ between 5 and 20, including 5 and 20, we can construct $E_\lambda$ using (i) of Theorem 12. Likewise, (ii) of Theorem 11 allows construction of $E_\lambda$ for all even integers between 2 and 30, including 30. We also have the choice of $E_{30} := M_0(3, 2, 2, 1)$ or $E_{30} := 2D_6$. Letting $a = 1$, $m = 3$, $u = 5$ in (v) of Theorem 12, we can construct an $E_{21}$; for example,

$$
E_{21} := \begin{bmatrix} 0\,2\,2\,1\,2\,2 \\ 2\,0\,2\,1\,2\,2 \\ 2\,2\,0\,1\,2\,2 \\ 1\,1\,1\,0\,0\,0 \\ 2\,2\,2\,0\,0\,0 \\ 2\,2\,2\,0\,0\,0 \end{bmatrix}.
$$

Assertion 3 of the Corollary to Theorem 10 permits us to take

$$E_{23} := E_{(3,6)}(6; 1, 1, 1), \quad E_{25} := E_{(2,5)}(6; 1, 1, 1), \quad E_{27} := E_{(2,3)}(6; 1, 1, 1).$$

It appears that our theorems *do not* provide means to construct an $E_3$ or an $E_{29}$. By ad hoc methods, we find:

$$E_3 := \begin{bmatrix} 0 & 2 & 0 & 0 & 0 & 1 \\ 2 & 0 & 0 & 0 & 0 & 0 \\ 0 & 0 & 0 & 0 & 0 & 0 \\ 0 & 0 & 0 & 0 & 0 & 0 \\ 0 & 0 & 0 & 0 & 0 & 0 \\ 1 & 0 & 0 & 0 & 0 & 0 \end{bmatrix}, \quad E_{29} := \begin{bmatrix} 0 & 2 & 2 & 2 & 2 & 1 \\ 2 & 0 & 4 & 0 & 0 & 4 \\ 2 & 4 & 0 & 4 & 0 & 0 \\ 2 & 0 & 4 & 0 & 4 & 0 \\ 2 & 0 & 0 & 4 & 0 & 4 \\ 1 & 4 & 0 & 0 & 4 & 0 \end{bmatrix}.$$

2. If $(N, d) = (8, 15)$, then letting $(b, m) = (1, 3)$, $w = 1, 2$ in (iv) of Theorem 12 yields $E_\lambda \in E(8, \le 15)$, with $2\lambda = \|E_\lambda\|$ and $Symm_8(\delta(z, E_\lambda)) \ne 0$ for $\lambda = 47, 49$. Likewise, if $(N, d) = (10, 35)$, then letting $(b, m) = (2, 3)$ and choosing a $w$ such that $1 \le w \le 10$ in (iv) of Theorem 11, we obtain $E_\lambda \in E(10, \le 35)$, with $2\lambda = \|E_\lambda\|$ and $Symm_{10}(\delta(z, E_\lambda)) \ne 0$ for values of $\lambda$ in the sequence $107, 109, \ldots, 125$.

3. Let $(N, d) = (8, 14)$. For $38 \le \lambda \le 54$, employing the Corollary of Theorem 10 with $a = 1$, we can construct an admissible matrix $A$ having either 1, 2 or 3 rows such that its corresponding $E$ satisfies $\|2D_N - E\| = 2\lambda$. Moreover, each of these $2D_N - E$ belongs to $E(8, 14)$ and the referred Corollary ensures that $Symm_8(\delta(z, E_\lambda)) \ne 0$.

4. Theorem 9 can be used to construct skew binary invariants (usually for large $N$), e.g., let $E \in E(26, 45)$ be the block-matrix

$$E := \begin{bmatrix} 0 & C \\ C^T & M(3, 5, 2, 1) \end{bmatrix},$$

where the $11 \times 15$ matrix $C := [c_{ij}]$ has $c_{ij} = 3$ for all $(i, j)$. Then, it is easily verified that Theorem 9 is applicable and thus $Symm_{26}(\delta(z, E))$ is a nonzero binary invariant of weight 585.

5. The skew binary invariant of least weight obtained from the Corollary to Theorem 12 corresponds to $\alpha := 5$, $\beta := 7$ and $\gamma := 3$. An easy example (which can also be thought of as an application of Theorem 9) is $I_{105}$. Here, $E \in E(10, 21)$ is the block-matrix

$$E := \begin{bmatrix} 2D_7 & C \\ C^T & 0 \end{bmatrix},$$

where the $7 \times 3$ matrix $C$ may be taken to have each entry equal to 3. As a SAGE computation shows, even after discounting permutations of rows and columns, there are 3719 choices of such $C$. On the other hand, $\text{dinv}_k(10, 21) = 547$. Thus, one is tempted to ask: does the set of all $Symm_{10}(\delta(z, E))$, as $E$ varies corresponding to these 3719 possible choices of $C$, generate the vector space $Inv_k(10, 21)$?

6. Our theorems do not allow construction of an $E \in E(5, 18)$ such that

$$Symm_5(\delta(z, E)) \neq 0.$$

In other words, Hermite's quintic skew invariant of weight 45 (which is unique up to a scalar multiple on account of the fact that $\text{dinv}_k(5, 18) = 1$) cannot be constructed using Theorems 7–12. Let $E_1$, $E_2 \in E(5, 18)$ be the matrices in block-format defined by

$$E_j := \begin{bmatrix} 0 & A_j \\ A_j^T & B \end{bmatrix}, \quad \text{where } j = 1, 2 \quad B := \begin{bmatrix} 0 & 1 & 7 \\ 1 & 0 & 1 \\ 7 & 1 & 0 \end{bmatrix},$$

$$A_1 := \begin{bmatrix} 5 & 13 & 0 \\ 5 & 3 & 10 \end{bmatrix} \quad \text{and} \quad A_2 := \begin{bmatrix} 8 & 10 & 0 \\ 2 & 6 & 10 \end{bmatrix}.$$

Then, a MAPLE computation shows that

$$Symm_5(\delta(z, E_1)) \neq 0 \neq Symm_5(\delta(z, E_2)).$$

Hence $E_1$, $E_2$ each yields the aforementioned Hermite's invariant. Thanks to Brendan McKay's *gtools suit*, not only does it tell us that there are 664 essentially distinct (i.e., non-conjugate under $S_5$) such matrices $E$, but it lists all of them in a few seconds. Using this list, a MAPLE computation demonstrates that $Symm_5(\delta(z, E)) \neq 0$ for exactly 223 matrices $E$ in the list. In other words, there are 223 pairwise nonisomorphic 18-regular (loopless, multi-) graphs on 5 vertices whose corresponding symmetrized-graph-monomials yield Hermite's quintic skew invariant. For an in-depth geometric study of this interesting invariant, the reader is referred to [15].

**Definitions** Let $E := [\varepsilon(i, j)]$ be an $N \times N$ symmetric matrix with nonnegative integer entries and having 0 principle diagonal, i.e., $E \in E(N)$.

1. Let *support of E* be the set

$$suppt(E) := \{\varepsilon(i, j) \mid 1 \leq i < j \leq N\}.$$

2. Let *bound of E* be the nonnegative integer

$$bound(E) := \max suppt(E).$$

3. For $\sigma \in S_N$, let $P_\sigma$ denote the $N \times N$ permutation matrix associated with $\sigma$, i.e., the $N \times N$ matrix whose $i$th row is the $\sigma(i)$th row of the $N \times N$ identity matrix.

Define
$$Symgrp(E) := \{\sigma \in S_N \mid P_\sigma E P_\sigma^T = E\} \quad \text{and}$$
$$symgrp(E) := \{\sigma \in S_N \mid \sigma(\delta(z, E)) = \delta(z, E)\}.$$

*Remarks*

1. Note that $symgrp(E)$ is a subgroup of $Symgrp(E)$. In the special case where each entry of $E$ is an even integer, $symgrp(E) = Symgrp(E)$.
2. For $E \in E(N)$, the full symmetrization of $\delta(z, E)$ is obtained by summing $\sigma(\delta(z, E))$ as $\sigma$ runs over a complete set of representatives of the (left) cosets of $symgrp(E)$ in $S_N$ and then multiplying the resulting sum by the integer $|symgrp(E)|$; an obvious advantage of an $E$ with a large $symgrp(E)$ is that there are fewer summands to deal with in the computation of $Symm_N(\delta(z, E))$.

**Definitions** Let $m, n$ be positive integers.

1. Given an $n \times n$ matrix $A$, let $D_m(A)$ denote the $m \times m$ block-matrix $[C_{ij}]$, where each $C_{ij}$ is an $n \times n$ matrix, $C_{ii} = 0$ for $1 \leq i \leq m$, and $C_{ij} = A = C_{ji}^T$ for $1 \leq i < j \leq m$.
2. Given $n \times n$ matrices $A_1, \ldots, A_m$, let $diag(A_1, A_2, \ldots, A_m)$ denote the $m \times m$ block-matrix $[C_{ij}]$ such that each $C_{ij}$ is an $n \times n$ matrix, $C_{ii} = A_i$ for $1 \leq i \leq m$, and $C_{ij} = 0 = C_{ji}$ for $1 \leq i < j \leq m$.

*Remarks* Assume that $m \geq 2$ and let $N := mn$.

1. The $N \times N$ matrix $diag(A_1, A_2, \ldots, A_m)$ is symmetric if and only if $A_i$ is symmetric for $1 \leq i \leq m$.
2. The $N \times N$ matrix $diag(A_1, A_2, \ldots, A_m)$ is a permutation matrix if and only if $A_i$ is an $n \times n$ permutation matrix for $1 \leq i \leq m$.
3. Let $\theta \in S_N$ be the permutation defined by $\theta(i) := l + 1 + (r - 1)n$, provided $i = lm + r$ with $0 \leq l \leq n - 1$ and $1 \leq r \leq m$. For $1 \leq r \leq m$, define

$$A_r = \{(r - 1)n + l + 1 \mid 0 \leq l \leq n - 1\} = \{\theta(lm + r) \mid 0 \leq l \leq n - 1\}.$$

Let $[a(i, j)] := M(m, n, b/2, c)$. Define $\varepsilon$ as follows:

$$\varepsilon(j, i) = \varepsilon(i, j) := \begin{cases} 0 & \text{if } 1 \leq i = j \leq N, \\ a(\theta^{-1}(i),\ \theta^{-1}(j)) & \text{if } (i, j) \in \pi[N]. \end{cases}$$

Observe that for $(i, j) \in \pi[N]$, we have

$$\varepsilon(i, j) = 0 \quad \text{if and only if} \quad (i, j) \in \pi(A_r) \quad \text{with } 1 \leq r \leq m.$$

Furthermore, given $1 \leq r < s \leq m$ and $(i, j) \in A_r \times A_s$, letting $i = (r - 1)n + l_1 + 1, j = (s - 1)n + l_2 + 1$ with $0 \leq l_1, l_2 \leq n - 1$, we have

$$a(\theta^{-1}(i),\ \theta^{-1}(j)) = a(l_1 m + r, l_2 m + s) = \begin{cases} b & \text{if } l_1 = l_2, \\ c & \text{if } l_1 \neq l_2, \end{cases}$$

and hence $[\varepsilon(i,j)] = D_m(cD_n + bI)$. Thus we get an $N \times N$ permutation matrix $P$ such that

$$M(m, n, b/2, c) = P \cdot D_m(cD_n + bI) \cdot P^{-1}.$$

In particular,

$$Symm_N(\delta(z, D_m(cD_n + bI))) = Symm_N(\delta(z, M(m, n, b/2, c))).$$

4. Suppose $m \geq 2$, $n = 2$ and that the entries of the $2 \times 2$ matrix $A$ are nonnegative integers. Further, assume that each row-sum of $D_m(A)$ as well as each column-sum of $D_m(A)$ is $d(m-1)$, where $d$ is a nonnegative integer. Then it is easy to verify that $A = (d-r)D_2 + rI$ with $0 \leq r \leq d$.

5. For a nonnegative integer $a$, we have

$$Symm_N(\delta(z, D_m((2a+1)I))) = 0.$$

This follows from the fact, seen above, that $D_m((2a+1)I)$ is a permutation-conjugate of $M(m, n, (2a+1)/2, 0)$.

**Theorem 13** *Let $m$, $n$ be positive integers and $N := mn$. Let $A$ and $B$ be $n \times n$ matrices. Suppose $Q$ is an $N \times N$ permutation matrix such that $QD_m(A)Q^T = D_m(B)$. Express $Q$ as an $m \times m$ block-matrix $[C_{ij}]$, where each $C_{ij}$ has size $n \times n$. Then the following holds.*

(i) *Assume that $A := [a_{ij}]$, where $a_{ij} \neq 0$ for all $(i,j) \in \{1, 2, \ldots, n\}^2$. Then, there exists a permutation $\sigma \in S_m$ such that $C_{ij} = 0$ whenever $j \neq \sigma(i)$ and $C_{i\sigma(i)}$ is an $n \times n$ permutation matrix for $1 \leq i \leq m$.*

(ii) *Assume $m \geq 3$ and that $A$ is symmetric and does not contain two identical columns. Also, assume that for some permutation $\sigma \in S_m$,*

$$C_{ij} = \begin{cases} P_i & \text{if } j = \sigma(i), \\ 0 & \text{if } j \neq \sigma(i), \end{cases}$$

*where $P_i$ is an $n \times n$ permutation matrix for $1 \leq i \leq m$. Then $P_i = P_j$ for $1 \leq i < j \leq m$.*

(iii) *Assume that $A = sD_n + rI$, where $r$ and $s$ are nonzero real numbers such that $r \neq s$. Then, $Symgrp(D_m(A)) \cong S_m \times S_n$. Furthermore, if $r$, $s$ are nonzero integers such that $r$ is even, then*

$$symgrp(D_m(A)) = Symgrp(D_m(A)).$$

*Proof* Let $\theta \in S_N$ be the permutation whose matrix is $Q$, i.e., the $ij$th entry of $QD_m(A)Q^T$ equals the $\theta(i)\theta(j)$th entry of $D_m(A)$. Let $i := qn + r$ and $j := qn + s$, where $0 \leq q \leq m - 1$ and $1 \leq r$, $s \leq n$. Then the $ij$th entry of $D_m(B)$ is 0 and hence the $\theta(i)\theta(j)$th entry of $D_m(A)$ must also be 0. Now the hypothesis of (i) implies $\theta(i) = ln + r_1$ as well as $\theta(j) = ln + s_1$ for some $0 \leq l \leq m - 1$ and some $1 \leq r_1$, $s_1 \leq n$. In

other words, letting $L(q) := \{qn + r \mid 1 \leq r \leq n\}$ for $0 \leq q \leq m - 1$, our $\theta$ induces
a permutation $\sigma$ of $\{0, \ldots, m - 1\}$ characterized by the property: $\theta(L(q)) = L(\sigma(q))$.
Now assertion (i) readily follows.

From the hypotheses of (ii), it is straightforward to deduce that if $P$ is an $n \times n$
permutation matrix with either $AP = A$ or $PA = A$, then $P$ is the identity matrix.
Since $A = A^T$ by hypothesis, we have $P_i A P_j^T = B$ for $1 \leq i < j \leq m$. In particular,
$P_1 A P_r = P_1 A P_s$ for $2 \leq r < s \leq m$ and hence $P_r = P_s$ for $2 \leq r < s \leq m$. Finally,
using the fact that $P_1 A P_3 = P_2 A P_3$, we get $P_1 = P_2$. Thus (ii) holds.

Assertion (iii) is obvious if $m = 1$. Henceforth, assume $m \geq 2$. For any $n \times n$
permutation matrix $P$, we clearly have $PD_nP^T = D_n$ and hence $PAP^T = A$. If $P_1$ and
$P_2$ are $n \times n$ permutation matrices with $P_1 A P_2^T = A$, then $A P_1 P_2^T = A$. Since no two
columns of $A$ are identical, we must have $P_1 = P_2$. Next, given $\theta \in Symgrp(D_m(A))$,
apply (i) with $B := A$ and $Q := P_\theta$ to deduce the existence of an ordered pair $(\sigma, \tau) \in$
$S_m \times S_n$ (induced by $\theta$) such that $Q$ is the $m \times m$ block matrix $[C_{ij}]$, where

$$C_{ij} = \begin{cases} P_\tau & \text{if } j = \sigma(i), \\ 0 & \text{if } j \neq \sigma(i). \end{cases}$$

This yields an injective group-homomorphism $Symgrp(D_m(A)) \to S_m \times S_n$. The fact
that this homomorphism is also surjective is straightforward to verify. Arguments
employed in the above proof of (i) allow us to identify a $\theta \in Symgrp(D_m(A))$ as an
ordered pair $(\sigma, \tau)$, where $\sigma$ is a permutation of $\{0, 1, \ldots, m - 1\}$, $\tau$ is a permutation
of $\{1, \ldots, n\}$, and if $i := qn + u$ with

$$(q, u) \in \{0, 1, \ldots, m - 1\} \times \{1, \ldots, n\},$$

then $\theta(i) = \sigma(q)n + \tau(u)$. Now suppose $r$ is a nonzero even integer and $s$ is a nonzero
integer such that $r \neq s$. For $0 \leq q_1 < q_2 \leq m - 1$, define $J(q_1 < q_2)$ to be the set
of all ordered pairs $(i, j)$ such that $1 \leq (i - q_1n) \leq n$, $1 \leq (j - q_2n) \leq n$ and $(i - q_1n) \neq (j - q_2n)$. Define $\Lambda$ to be the set of all $(i, j) \in \pi(N)$ such that $i \equiv j \bmod n$.
Then $J(q_1 < q_2) \subset \pi(N)$ and

$$\delta(z, D_m(A)) = \prod_{(i,j)\in\Lambda} (z_i - z_j)^r \prod_{0\leq q_1<q_2\leq m-1} \prod_{(i,j)\in J(q_1<q_2)} (z_i - z_j)^s.$$

Fix a $\theta := (\sigma, \tau)$ in $Symgrp(D_m(A))$. Since $\theta$ permutes $\Lambda$ and $r$ is an even integer,

$$\prod_{(i,j)\in\Lambda} (z_{\theta(i)} - z_{\theta(j)})^r = \prod_{(i,j)\in\Lambda} (z_i - z_j)^r.$$

Given an $(i, j) \in J(q_1 < q_2)$, observe that $\theta(i) > \theta(j)$ if and only if $\sigma(q_1) > \sigma(q_2)$.
So, if $\sigma(q_1) < \sigma(q_2)$, then $\theta(i) < \theta(j)$ for all $(i, j) \in J(q_1 < q_2)$ and it follows that

$$\prod_{(i,j)\in J(q_1<q_2)} (z_{\theta(i)} - z_{\theta(j)})^s = \prod_{(i,j)\in J(\sigma(q_1)<\sigma(q_2))} (z_i - z_j)^s.$$

On the other hand, if $\sigma(q_1) > \sigma(q_2)$, then $\theta(i) > \theta(j)$ for all $(i,j) \in J(q_1 < q_2)$ and then

$$\prod_{(i,j)\in J(q_1<q_2)} (z_{\theta(i)} - z_{\theta(j)})^s = (-1)^{n(n-1)s} \prod_{(i,j)\in J(\sigma(q_2)<\sigma(q_1))} (z_i - z_j)^s.$$

Since $n(n-1)s$ an even integer, we conclude that $\theta(\delta(z, D_m(A))) = \delta(z, D_m(A))$. Assertion (iii) is thus fully established. $\qquad\square$

## 2.4 Fermions in the $\nu < 1/2$ IQL State

Let $N$, $n$ be positive integers such that $N \geq 3$. Let $\nu$ be a rational number of the form $n/(2pn \pm 1)$, where $p$ is a positive integer. It is tacitly assumed that $\nu < 1/2$, i.e., either $p \geq 2$ or $\nu \neq n/(2n-1)$. In this section (as in [10]), we present several configurations, including the minimal configurations, of $N$ Fermions in the IQL state with filling factor $\nu$. Given a filling factor $n/(2pn \pm 1)$, we let $m$ denote the integer $N/n$. Note that, necessarily, we have $m \geq 2$. For each configuration of $N$ Fermions in the IQL state with filling factor $\nu$, the corresponding correlation function $G(z_1, \ldots, z_N)$ that we construct is a nonzero homogeneous polynomial. Since the total angular momentum of such a system is 0, it is mandated that the total degree of $G(z_1, \ldots, z_N)$ be

$$\kappa_G := N\ell - \frac{N(N-1)}{2} = \frac{1}{2}N[(2p-1)n \pm 1](m-1).$$

It is interesting to observe that for all $N$ and $\nu$, the corresponding number $\kappa_G$ is an even integer. For each configuration, we obtain the corresponding $G(z_1, \ldots, z_N)$ by symmetrizing a suitable $\delta(z, E)$, where the matrix $E$ is in $E(N, 2\ell - N + 1)$. A noteworthy consequence of our construction is that by (i) of Theorem 5, the correlation polynomial $G(z_1, \ldots, z_N)$ is a binary invariant of type $(N, 2\ell - N + 1)$, and its $z_i$-degree equals $2\ell - N + 1$ for $1 \leq i \leq N$. From the above mentioned fact that $G(z_1, \ldots, z_N)$ has even degree, i.e., $\kappa_G$ is even, and the fact that $G(z_1, \ldots, z_N)$ is homogeneous, we deduce the additional property:

$$G(-z_1, \ldots, -z_N) = G(z_1, \ldots, z_N).$$

The energy of an IQL configuration is directly related to the pair correlations; so, it is useful to keep track of the proportion of Fermion-pairs with a given correlation potency. For this purpose, given a configuration with its associated $E := [\varepsilon(i,j)] \in E(N)$ and given an integer $b$, let *frequency of $b$ in $E$*, denoted by $frq(b, E)$, be the number of pairs with correlation-potency $b$, i.e.,

$$frq(b, E) = |\{(i,j) \mid 1 \leq i < j \leq N \text{ with } \varepsilon(i,j) = b\}|.$$

The aforementioned proportion is the ratio of $frq(b, E)$ to $N(N - 1)/2$. Heuristically, the limit of each proportion as $N$ increases to infinity provides an insight into the qualitative nature of the configuration; in particular, it is a distinguishing feature of the minimal configurations. It is worth pointing out that if a configuration presented below is admissible for all $n$, then it specializes to the Laughlin configuration when $n = 1$, i.e., when $\nu = 1/(2p \pm 1)$.

### 2.4.1  $\nu = n/(2pn + 1)$

Assume $\nu = n/(2pn + 1)$. Then we have

$$2\ell = \nu^{-1}N - (2pn + 1) - 1 + n.$$

Consequently,

$$2\ell - N + 1 = [(2p - 1)n + 1](m - 1).$$

Let $G := Symm_N(\delta(z, E))$, where $E := M(m, n, p, 2p - 1)$. Assertion (ii) of the Corollary of Theorem 8 ensures that $G(z_1, \ldots, z_N)$ is a nonzero polynomial. $G(z_1, \ldots, z_N)$ is homogeneous of total degree

$$\kappa_G := Nl - \frac{N(N - 1)}{2} = \frac{1}{2}N[(2p - 1)n + 1](m - 1),$$

and its $z_i$-degree is $2\ell - N + 1$ for $1 \le i \le N$. Moreover, by (iii) of Theorem 13, $symgrp(E)$ is isomorphic to $S_m \times S_n$. From the definition of $M(m, n, p, 2p - 1)$, it is clear that $suppt(E) = \{0, 2p - 1, 2p\}$,

$$frq(b, E) = \begin{cases} \frac{N(n-1)}{2} & \text{if } b = 0, \\[2mm] \frac{N(N-n)(n-1)}{2n} & \text{if } b = 2p - 1, \\[2mm] \frac{N(N-n)}{2n} & \text{if } b = 2p. \end{cases}$$

This correlation-statistics is organized in Table 2.3. Specific cases when $n = 2, 3$ and $p = 1$ are illustrated in Fig. 2.6. Observe that if $n \ge 3$, then $E$ corresponds to the unique minimal configuration for the filling factor $n/(2pn + 1)$. If $n = 2$, then the only additional minimal configuration corresponds to $M(m, 2, p - (1/2), 2p)$; by (ii) of the Corollary of Theorem 8, this is an existent configuration; its correlation-statistics is identical to the one presented in Table 2.3.

From the correlation-statistics, it at once follows that as $N$ increases to $\infty$, the proportion of uncorrelated Fermion-pairs tends to 0, the proportion of $(2p - 1)$-correlated Fermion-pairs tends to $(n - 1)/n$ and the proportion of $(2p)$-correlated Fermion-pairs tends to $1/n$.

**Table 2.3** Correlation-statistics; IQL state, $\nu = n/(2pn + 1)$

| correl. potency | 0 | $2p - 1$ | $2p$ |
|---|---|---|---|
| No. of pairs | $\frac{N(n-1)}{2}$ | $\frac{N(n-1)(N-n)}{2n}$ | $\frac{N(N-n)}{2n}$ |
| Proportion | $\frac{n-1}{N-1}$ | $\frac{(n-1)(N-n)}{n(N-1)}$ | $\frac{N-n}{n(N-1)}$ |

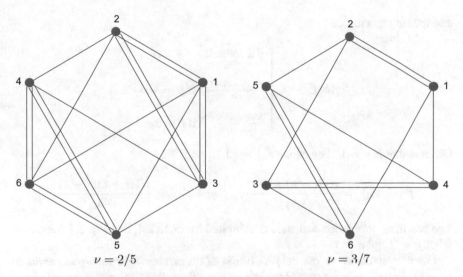

$\nu = 2/5$  $\nu = 3/7$

**Fig. 2.6** $N = 6$; IQL state, minimal configurations for filling factors $2/5$, $3/7$

We proceed to present some noteworthy configurations, which are non-minimal if $n \geq 3$, and which arise when $N$, $n$ and $p$ satisfy certain special properties. First, consider the special case where $m$ is an odd integer. Then $m \geq 3$ and hence we may let $G := Symm_N(\delta(z, E_0))$, where

$$E_0 := M_0(m, n, p, 2p - 1).$$

It can be verified that $symgrp(E_0)$ contains a subgroup isomorphic to the semi-direct product of $S_n$ and a cyclic group of order $m$. Assertion (ii) of Theorem 7 ensures that $G(z_1, \ldots, z_N)$ is a nonzero polynomial. Also, $G(z_1, \ldots, z_N)$ is homogeneous in $z_1, \ldots, z_N$ of total degree

$$\kappa_G := Nl - \frac{N(N-1)}{2} = \frac{1}{2}N[(2p-1)n + 1](m-1),$$

and its $z_i$-degree is $2\ell - N + 1$ for $1 \leq i \leq N$.

From the definition of $M_0(m, n, p, 2p - 1)$, it follows that

$$suppt(E_0) = \{0, 4p - 2, 2p\},$$

**Table 2.4** Correlation-statistics; IQL state $\nu = n/(2pn + 1)$ and $N/n$ odd

| correl. potency | 0 | $2(2p - 1)$ | $2p$ |
|---|---|---|---|
| No. of pairs | $\frac{N(N+n)(n-1)}{4n}$ | $\frac{N(N-n)(n-1)}{4n}$ | $\frac{N(N-n)}{2n}$ |
| Proportion | $\frac{(N+n)(n-1)}{2n(N-1)}$ | $\frac{(N-n)(n-1)}{2n(N-1)}$ | $\frac{N-n}{n(N-1)}$ |

and the frequencies are

$$
frq(b, E_0) = \begin{cases} \frac{N(N+n)(n-1)}{4n} & \text{if } b = 0, \\[2mm] \frac{N(N-n)(n-1)}{4n} & \text{if } b = 4p - 2, \\[2mm] \frac{N(N-n)}{2n} & \text{if } b = 2p. \end{cases}
$$

Observe that if $p = 1$, then $supp t(E_0) = \{0, 2\}$ and

$$
frq(2, E_0) = \frac{N(N - n)(n + 1)}{4n}, \quad frq(0, E_0) = \frac{N(N + n)(n - 1)}{4n}.
$$

The resulting correlation-statistics is exhibited in Table 2.4. See Fig. 2.7 for a case when $n = 2$ and $p = 1$.

Observe that as $N \to \infty$, the proportion of uncorrelated Fermion-pairs tends to $(n - 1)/2n$ (unlike in the case of the minimal configuration), the proportion of $(4p - 2)$-correlated Fermion-pairs also tends to $(n - 1)/2n$, whereas the proportion of $(2p)$-correlated Fermion-pairs tends to $1/n$ (as in the case of the minimal configuration).

Next, consider the case where $n$ is an odd integer $\geq 3$ and $2p$ is an integer multiple of $n - 1$ (e.g., $n = 3$); say $2p = s(n - 1)$. Let $G := Symm_N(\delta(z, E_0))$, where $E_0 := M_0(n, m, p(m - 1), s - 1)$. As before, $symgrp(E_0)$ contains a subgroup isomorphic to the semi-direct product of $S_n$ and a cyclic group of order $m$. Assertion (ii) of Theorem 7 ensures that $G(z_1, \ldots, z_N)$ is a nonzero polynomial. Also, $G(z_1, \ldots, z_N)$ is homogeneous in $z_1, \ldots, z_N$ of total degree

$$
\kappa_G := Nl - \frac{N(N - 1)}{2} = \frac{1}{2}N[(2p - 1)n + 1](m - 1),
$$

and its $z_i$-degree is $2\ell - N + 1$ for $1 \leq i \leq N$. It readily follows that

$$
supp t(E_0) = \begin{cases} \{0, \; 2p(m - 1), \; 2(s - 1)\} & \text{if } s \geq 2, \\ \{0, \; 2p(m - 1)\} & \text{if } s = 1. \end{cases}
$$

**Fig. 2.7** $N = 6$; IQL state for $\nu = 2/5$, a non-minimal configuration

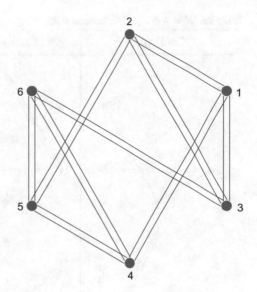

**Table 2.5** Correlation-stats.; IQL state $\nu = n/(sn(n-1)+1)$ with $n$ odd $\geq 3$ and $s \geq 1$

| $s$ | $\geq 2$ | | | 1 | |
|---|---|---|---|---|---|
| correl. potency | 0 | $2p(m-1)$ | $2(s-1)$ | 0 | $2p(m-1)$ |
| No. of pairs | $\frac{N(N-n)(n+1)}{4n}$ | $\frac{N(n-1)}{2}$ | $\frac{N(N-n)(n-1)}{4n}$ | $\frac{N(N-n)}{2}$ | $\frac{N(n-1)}{2}$ |
| Proportion | $\frac{(N-n)(n+1)}{2n(N-1)}$ | $\frac{n-1}{N-1}$ | $\frac{(N-n)(n-1)}{2n(N-1)}$ | $\frac{N-n}{N-1}$ | $\frac{n-1}{N-1}$ |

Moreover, the frequencies are:

$$
frq(b, E_0) = \begin{cases}
\frac{N(N-n)(n+1)}{4n} & \text{if } b = 0, s \geq 2, \\
\frac{N(N-n)}{2} & \text{if } b = 0, s = 1, \\
\frac{N(n-1)}{2} & \text{if } b = 2p(m-1), \\
\frac{N(N-n)(n-1)}{4n} & \text{if } b = 2(s-1) \text{ with } s \geq 2, \\
0 & \text{if } b = 2(s-1) \text{ with } s = 1.
\end{cases}
$$

Thus, we obtain Table 2.5 which tabulates the resulting correlation-statistics.

If $s = 1$, then the proportion of uncorrelated Fermion-pairs tends to 1 as $N \to \infty$. If $s \geq 2$, then as $N \to \infty$, the proportion of uncorrelated Fermion-pairs tends to $(n+1)/2n$, the proportion of $(2pm - 2p)$-correlated Fermion-pairs tends to 0 and the proportion of $(2s - 2)$-correlated Fermion pairs tends to $(n-1)/2n$. So, in comparison with the minimal configuration, the distribution of correlations is non-uniform. The leftmost configuration in the second row of Table 2.6 is a configuration of this type for $N = 6$ and $\nu = 3/7$.

**Table 2.6** $N = 6$ & $\nu = 3/7$; existent IQL configurations with bound 2

| Configs. I | Configs. II | Configs. III |
|---|---|---|

**Fig. 2.8** $N = 6$ & $\nu = 3/7$; IQL state, the nonexistent config. of bound 2

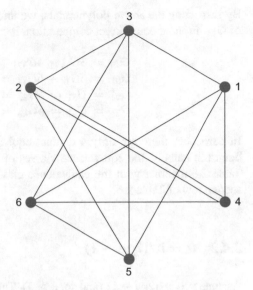

As an example, we proceed to work out the case of $N = 6$ with $\nu = 3/7$ in a comprehensive manner. In this case, there are exactly 24 apparent configurations whose corresponding diagrams are 4-regular multi-graphs on 6 vertices. It is easy to see that the least possible *bound* for any configuration of this kind is 2. Out of the 24 apparent configurations, exactly 13 have this least bound. Further, out of the 13 apparent configurations with the least bound, exactly 12 are existent; they are presented in Table 2.6. The only nonexistent configuration is the one in Fig. 2.8.

Regard Table 2.6 as a $4 \times 3$ matrix of diagrams. Observe that the figure at position $(1, 1)$ is the unique minimal configuration. Let $G_{ij} := G_{ij}(z_1, \ldots, z_6)$ denote the correlation function associated with the diagram in the $i$th row, $j$th column of Table 2.6. For the sake of compactness, we prefer to express the functions $G_{ij}$ as polynomials in $y_1, \ldots, y_5$, where $y_r$ denotes the coefficient of $X^{5-r}$ in

$$\prod_{j=1}^{6} \left( X + z_j - \frac{1}{6}(z_1 + z_2 + z_3 + z_4 + z_5 + z_6) \right)$$

for $1 \le r \le 5$. A direct computation shows that

$$\begin{aligned}
G_{11} = {}& -1536\, y_5 y_1{}^3 + 384\, y_1{}^2 y_2 y_4 + 384\, y_1{}^2 y_3{}^2 - 384\, y_1 y_2{}^2 y_3 \\
& + 72\, y_2{}^4 + 31488\, y_1 y_3 y_5 - 9600\, y_1 y_4{}^2 - 16128\, y_5 y_2{}^2 \\
& + 5760\, y_2 y_3 y_4 - 1536\, y_3{}^3 + 63360\, y_5{}^2,
\end{aligned}$$

$$\begin{aligned}
G_{12} = {}& -960\, y_5 y_1{}^3 + 240\, y_1{}^2 y_2 y_4 + 432\, y_1{}^2 y_3{}^2 - 384\, y_1 y_2{}^2 y_3 \\
& + 72\, y_2{}^4 + 25440\, y_1 y_3 y_5 - 6000\, y_1 y_4{}^2 - 12240\, y_5 y_2{}^2 \\
& + 3600\, y_2 y_3 y_4 - 960\, y_3{}^3 + 82800\, y_5{}^2.
\end{aligned}$$

By inspecting the above polynomials, we find that $G_{12}$ is not a numerical multiple of $G_{11}$. From another direct computation, it can be verified that

$$
\begin{aligned}
G_{13} &= 3G_{11} - 4G_{12}, \\
G_{21} &= 6G_{12} - 6G_{11} = 2G_{22} = 3G_{23}, \\
G_{21} &= G_{31} = 6G_{32} = G_{33}, \\
G_{21} &= G_{41} = 6G_{42} = 6G_{43}.
\end{aligned}
$$

In particular, there are only 4 distinct equivalence classes of configurations with bound 2; namely, the equivalence classes of the configurations in the first row of Table 2.6 together with the equivalence class of the leftmost configuration in the second row of Table 2.6.

### 2.4.2  $\nu = n/(2pn - 1)$

Assume $\nu = n/(2pn - 1)$ (and so, $p \geq 2$). Then, the individual angular momentum $\ell$ is the half-integer determined by the equation

$$
2\ell = \nu^{-1}N - (2pn - 1) - 1 + n.
$$

Hence, we clearly have

$$
2\ell - N + 1 = [(2p - 1)n - 1](m - 1).
$$

Let $G := Symm_N(\delta(z, E))$, where $E := M(m, n, p - 1, 2p - 1)$. Observe that $symgrp(E)$ is isomorphic to $S_m \times S_n$. Assertion (ii) of the Corollary of Theorem 8 assures that $G(z_1, \ldots, z_N)$ is a nonzero polynomial. Furthermore, $G(z_1, \ldots, z_N)$ is homogeneous in $z_1, \ldots, z_N$ of total degree

$$
\kappa_G := N\ell - \frac{N(N - 1)}{2} = \frac{1}{2}N[(2p - 1)n - 1](m - 1),
$$

and the $z_i$-degree of $G(z_1, \ldots, z_N)$ is $2\ell - N + 1$ for $1 \leq i \leq N$. From the definition of $M(m, n, p - 1, 2p - 1)$, it is clear that $suppt(E) = \{0, 2p - 2, 2p - 1\}$ and

$$
frq(b, E) = \begin{cases} \frac{N(n-1)}{2} & \text{if } b = 0, \\ \frac{N(N-n)}{2n} & \text{if } b = 2p - 2, \\ \frac{N(N-n)(n-1)}{2n} & \text{if } b = 2p - 1. \end{cases}
$$

If $n \geq 3$, then $M(m, n, p - 1, 2p - 1)$ corresponds to the unique minimal configuration for the filling factor $n/(2pn - 1)$. If $n = 2$, then the only additional minimal configuration corresponds to $M(m, 2, p - (1/2), 2(p - 1))$; by (ii) of the Corollary

**Table 2.7** Correlation-statistics; IQL state $\nu = n/(2pn - 1)$

| correl. potency | 0 | $2(p-1)$ | $2p-1$ |
|---|---|---|---|
| No. of pairs | $\frac{N(n-1)}{2}$ | $\frac{N(N-n)}{2n}$ | $\frac{N(N-n)(n-1)}{2n}$ |
| Proportion | $\frac{n-1}{N-1}$ | $\frac{N-n}{n(N-1)}$ | $\frac{(N-n)(n-1)}{n(N-1)}$ |

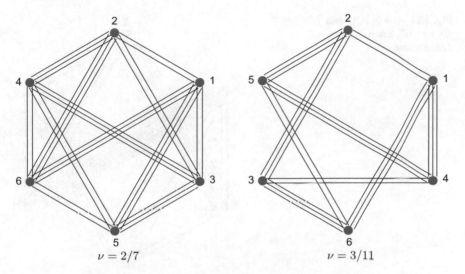

$\nu = 2/7$           $\nu = 3/11$

**Fig. 2.9** $N = 6$; IQL state, minimal configurations for filling factors $2/7, 3/11$

of Theorem 8, this is an existent configuration having the same correlation-statistics as that of $M(m, 2, p-1, 2p-1)$. Table 2.7 gives the correlation-statistics for $n \geq 2$. See Fig. 2.9 for particular configurations when $n = 2$ and $n = 3$.

As in the case of the minimal configuration for $\nu = n/(2pn + 1)$, the proportion of uncorrelated Fermion-pairs tends to 0 when $N \to \infty$. Also, as $N \to \infty$, the proportion of $(2p-2)$-correlated Fermion-pairs tends to $1/n$ and the proportion of $(2p-1)$-correlated Fermion-pairs tends to $(n-1)/n$. So, asymptotically, the odd correlations dominate over the even correlations when $n \geq 2$.

Next, consider the case where $m$ is an odd integer. Of course, we must have $m \geq 3$. In this case, we let $G := Symm_N(\delta(z, E_0))$, where

$$E_0 := M_0(m, n, p-1, 2p-1).$$

Here, we are content to note that $symgrp(E_0)$ contains a subgroup isomorphic to the semi-direct product of $S_n$ and a cyclic group of order $m$. Thanks to assertion (ii) of Theorem 7, we are assured that $G(z_1, \ldots, z_N)$ is a nonzero polynomial. As needed, $G(z_1, \ldots, z_N)$ is homogeneous in $z_1, \ldots, z_N$ of total degree

$$\kappa_G := Nl - \frac{N(N-1)}{2} = \frac{1}{2}N[(2p-1)n - 1](m-1),$$

**Table 2.8** Correlation-statistics; IQL state $\nu = n/(2pn-1)$ with $N/n$ odd

| correl. potency | 0 | $2(p-1)$ | $2(2p-1)$ |
|---|---|---|---|
| No. of pairs | $\frac{N(N+n)(n-1)}{4n}$ | $\frac{N(N-n)}{2n}$ | $\frac{N(N-n)(n-1)}{4n}$ |
| Proportion | $\frac{(N+n)(n-1)}{2n(N-1)}$ | $\frac{N-n}{n(N-1)}$ | $\frac{(N-n)(n-1)}{2n(N-1)}$ |

**Fig. 2.10** $N = 6$; IQL state for $\nu = 2/7$, a non-minimal configuration

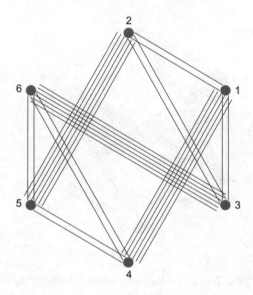

and the $z_i$-degree of $G(z_1, \ldots, z_N)$ is $2\ell - N + 1$ for $1 \le i \le N$. We have

$$suppt(E_0) = \{0, \ 2p-2, \ 4p-2\}.$$

The frequencies are described by

$$frq(b, E_0) = \begin{cases} \frac{N(N+n)(n-1)}{4n} & \text{if } b = 0, \\ \frac{N(N-n)}{2n} & \text{if } b = 2p-2, \\ \frac{N(N-n)(n-1)}{4n} & \text{if } b = 4p-2. \end{cases}$$

The correlation-statistics is presented in Table 2.8; see Fig. 2.10 for a specific example when $n = 2$.

In this configuration, as $N \to \infty$, the proportion of uncorrelated Fermion-pairs and the proportion of $(4p-2)$-correlated Fermion-pairs each tends to $(n-1)/2n$, but the proportion of $(2p-2)$-correlated Fermion-pairs tends to $1/n$.

Lastly, consider the case where $n \ge 3$ is odd and $2(p-1)$ is an integer multiple of $n-1$; say $2(p-1) = t(n-1)$ (e.g., $\nu = 3/(6p-1)$). As in the previous case, $symgrp(E_0)$ contains a subgroup isomorphic to the semi-direct product of $S_n$ and a cyclic group of order $m$. Let $G := Symm_N(\delta(z, E_0))$, where $E_0 := M_0(n, m, p(m-$

**Table 2.9** Correlation-stats.; IQL state $\nu = n/((tn - t + 2)n - 1)$ with $n \geq 3$ odd, $t \geq 1$

| $t$ | $\geq 2$ | | | 1 | |
|---|---|---|---|---|---|
| correl. potency | 0 | $2p(m-1)$ | $2(t-1)$ | 0 | $2p(m-1)$ |
| No. of pairs | $\frac{N(N-n)(n+1)}{4n}$ | $\frac{N(n-1)}{2}$ | $\frac{N(N-n)(n-1)}{4n}$ | $\frac{N(N-n)}{2}$ | $\frac{N(n-1)}{2}$ |
| Proportion | $\frac{(N-n)(n+1)}{2n(N-1)}$ | $\frac{n-1}{N-1}$ | $\frac{(N-n)(n-1)}{2n(N-1)}$ | $\frac{N-n}{N-1}$ | $\frac{n-1}{N-1}$ |

1), $t - 1$). Then, assertion (ii) of Theorem 7 ensures that $G(z_1, \ldots, z_N)$ is a nonzero polynomial. Also, $G(z_1, \ldots, z_N)$ is homogeneous in $z_1, \ldots, z_N$ of total degree

$$\kappa_G := Nl - \frac{N(N-1)}{2} = \frac{1}{2}N[(2p-1)n - 1](m-1),$$

and the $z_i$-degree of $G(z_1, \ldots, z_N)$ is $2\ell - N + 1$ for $1 \leq i \leq N$. We have

$$suppt(E_0) = \{0, 2p(m-1), 2(t-1)\}.$$

The frequencies are:

$$frq(b, E_0) = \begin{cases} \frac{N(N-n)(n+1)}{4n} & \text{if } b = 0 \text{ and } s \geq 2, \\ \frac{N(N-n)}{2} & \text{if } b = 0 \text{ and } t - 1, \\ \frac{N(n-1)}{2} & \text{if } b = 2p(m-1), \\ \frac{N(N-n)(n-1)}{4n} & \text{if } b = 2(t-1) \text{ and } t \geq 2. \end{cases}$$

Thus, we obtain the correlation-statistics as in Table 2.9; see Fig. 2.11 for a case when $n = 3$ and $t = 1$.

If $2p = n + 1$, then as $N \to \infty$, the proportion of uncorrelated Fermon-pairs tends to 1. If $2p \neq n + 1$, i.e., $t \geq 2$, then as $N \to \infty$, the proportion of uncorrelated Fermion-pairs tends to $(n + 1)/2n$, the proportion of $(2pm - 2p)$-correlated Fermion-pairs tends to 0 and the proportion of $(2t - 2)$-correlated Fermion-pairs tends to $(n - 1)/2n$. So, in either case, the proportion of uncorrelated pairs is the largest.

As an example, we provide a detailed account of the case of $N = 4$ with $\nu = 2/7$. There are only 5 apparent configurations to be considered in this case. The configurations presented in Figs. 2.12 and 2.13 are nonexistent. The only 2 existent configurations appear in Fig. 2.14.

By a direct computation, it is verified that the existent configurations yield the same correlation function $G(z_1, z_2, z_3, z_4)$. As before, for $r = 1, 2, 3$, let $y_r$ be the coefficient of $X^{3-r}$ in

$$\prod_{j=1}^{4}\left(X + z_j - \frac{1}{4}(z_1 + z_2 + z_3 + z_4)\right).$$

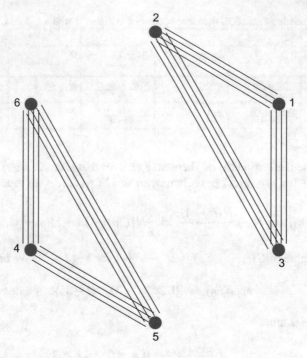

**Fig. 2.11** $N = 6$; IQL state for $\nu = 3/11$, a non-minimal configuration

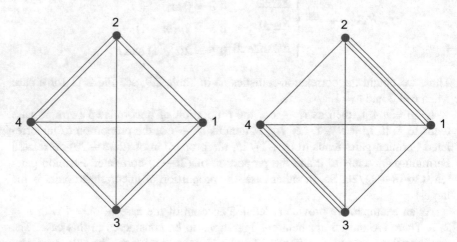

**Fig. 2.12** $N = 4$ & $\nu = 2/7$; nonexistent IQL configurations

Then $G := G(z_1, z_2, z_3, z_4)$ is compactly expressed as

$$G = 8 y_1{}^5 - 192 y_1{}^3 y_3 + 108 y_1{}^2 y_2{}^2 - 3456 y_1 y_3{}^2 + 1296 y_2{}^2 y_3.$$

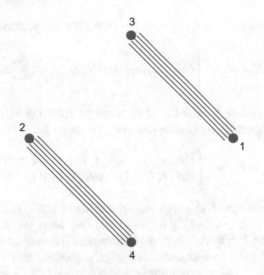

**Fig. 2.13** $N = 4$ & $\nu = 2/7$; a nonexistent IQL configuration

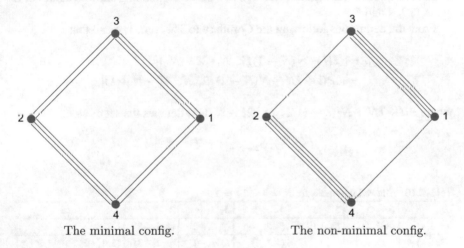

The minimal config.          The non-minimal config.

**Fig. 2.14** $N = 4$ & $\nu = 2/7$; the existent IQL configurations

## 2.5 Systems with QE in the $\nu = 1/3$ IQL

We begin by recalling basic facts about enumeration of distinct multiplets of a fixed total angular momentum $L$ in a system containing $N$ Fermions each having angular momentum $\ell$ (where $2\ell$ is a positive integer $\geq (N - 1)$). Such a system is represented as the collection $S(N, \ell)$ of *states*, where by a *state*, we mean an $N$-tuple $(\lambda_1, \ldots, \lambda_N)$ with

$$\lambda_i \in \{\ell - j \mid 0 \leq j \leq 2\ell\} \quad \text{for } 1 \leq i \leq N \text{ and } \quad \lambda_1 > \cdots > \lambda_N.$$

In particular, note that $\lambda_1 \le \ell$ and $\lambda_i + \ell - N + i$ is a nonnegative integer for $1 \le i \le N$. Let

$$S(N, \ell, t) := \left\{ (\lambda_1, \ldots, \lambda_N) \in S(N, \ell) \ \middle|\ t = \sum_i^N \lambda_i \right\}$$

and let $s(N, \ell, t)$ denote the number of elements (or states) in $S(N, \ell, t)$. Then, the *number of distinct multiplets of total angular momentum L* is

$$g(N, \ell, L) := \begin{cases} s(N, \ell, L) - s(N, \ell, L + 1) & \text{if } L \ge 0 \text{ and} \\ s(N, \ell, L + 1) - s(N, \ell, L) & \text{if } L < 0. \end{cases}$$

It is easily seen that $s(N, \ell, t) = s(N, \ell, -t)$ and hence if $L < 0$, then $g(N, \ell, L) = g(N, \ell, -L - 1)$ with $(-L - 1) \ge 0$. So, to find the values taken by $g(N, \ell, L)$, it suffices to assume $L \ge 0$. As an example, consider the case of $N = 3$ Fermions on the surface of a sphere with individual angular momentum $\ell = 3$. In this case, Table 2.10 exhibits the allowed multiplets, where $L_z$ stands for the spatial component of the total angular momentum. Observe that the corresponding allowed values of $L$ are 0, 2, 3, 4 and 6.

From the definitions following the Corollary to Theorem 1, recall that

$$p(t + \ell N - N(N - 1)/2, \ N, \ 2\ell - N + 1)$$
$$= |P(t + \ell N - N(N - 1)/2, \ N, \ 2\ell - N + 1)|,$$

where $P(t + \ell N - N(N - 1)/2, \ N, \ 2\ell - N + 1)$ denotes the set of all

$$(a_1, \ldots, a_{2\ell - N + 1}) \in \mathbb{N}^{2\ell - N + 1}$$

**Table 2.10** Allowed multiplets for $N = 3$ and $\ell = 3$

| $L_z$ | 6 | 5 | 4 | 3 | 2 | 1 | 0 |
|---|---|---|---|---|---|---|---|
| $(3, 2, *)$ | $(3, 2, 1)$ | $(3, 2, 0)$ | $(3, 2, -1)$ | $(3, 2, -2)$ | $(3, 2, -3)$ | | |
| $(3, 1, *)$ | | | $(3, 1, 0)$ | $(3, 1, -1)$ | $(3, 1, -2)$ | $(3, 1, -3)$ | |
| $(3, 0, *)$ | | | | | $(3, 0, -1)$ | $(3, 0, -2)$ | $(3, 0, -3)$ |
| $(3, -1, *)$ | | | | | | | $(3, -1, -2)$ |
| $(2, 1, *)$ | | | | $(2, 1, 0)$ | $(2, 1, -1)$ | $(2, 1, -2)$ | $(2, 1, -3)$ |
| $(2, 0, *)$ | | | | | | $(2, 0, -1)$ | $(2, 0, -2)$ |
| $(1, 0, *)$ | | | | | | | $(1, 0, -1)$ |
| $\lvert L, L_z \rangle$ | $\lvert 6, 6 \rangle$ | $\lvert 6, 5 \rangle$ | $\lvert 6, 4 \rangle$ | $\lvert 6, 3 \rangle$ | $\lvert 6, 2 \rangle$ | $\lvert 6, 1 \rangle$ | $\lvert 6, 0 \rangle$ |
| $\lvert L, L_z \rangle$ | | | $\lvert 4, 4 \rangle$ | $\lvert 4, 3 \rangle$ | $\lvert 4, 2 \rangle$ | $\lvert 4, 1 \rangle$ | $\lvert 4, 0 \rangle$ |
| $\lvert L, L_z \rangle$ | | | | $\lvert 3, 3 \rangle$ | $\lvert 3, 2 \rangle$ | $\lvert 3, 1 \rangle$ | $\lvert 3, 0 \rangle$ |
| $\lvert L, L_z \rangle$ | | | | | $\lvert 2, 2 \rangle$ | $\lvert 2, 1 \rangle$ | $\lvert 2, 0 \rangle$ |
| $\lvert L, L_z \rangle$ | | | | | | | $\lvert 0, 0 \rangle$ |

such that

$$\sum_{r=1}^{2\ell-N+1} a_r \leq N \quad \text{and} \quad \sum_{r=1}^{2\ell-N+1} ra_r = t + \ell N - \frac{N(N-1)}{2}.$$

The function

$$S(N, \ell, t) \rightarrow P(t + \ell N - N(N-1)/2, N, 2\ell - N + 1) \quad \text{given by}$$
$$(\lambda_1, \ldots, \lambda_N) \rightarrow (\lambda_1 + \ell - N + 1, \ldots, \lambda_i + \ell - N + i, \ldots, \lambda_N + \ell)$$

is easily verified to be a bijection. Consequently, we have

$$s(N, \ell, t) = p(t + \ell N - N(N-1)/2, N, 2\ell - N + 1).$$

Now recall that for positive integers $n$ and $d$,

$$\mathfrak{G}(n, d, X) := \frac{\prod_{i=1}^{d}(1 - X^{n+i})}{\prod_{i=2}^{d}(1 - X^i)}$$

is a polynomial in $X$ of degree $nd + 1$ and by the assertion (iv) of Theorem 5, $p(m, d, n) - p(m-1, d, n)$ is the coefficient of $X^m$ in $\mathfrak{G}(n, d, X)$. The well-known fact (see, e.g., [16, 17]) that for $L \geq 0$, $g(N, \ell, L)$ is the coefficient of $X^L$ in

$$(-1) \ X^{-1 - \frac{N(2\ell-N+1)}{2}} \cdot \mathfrak{G}(2\ell - N + 1, N, X) = X^{\frac{N(2\ell-N+1)}{2}} \cdot \mathfrak{G}(2\ell - N + 1, N, X^{-1})$$

follows readily. Let $q(n, d, X)$ and $r(n, d, X)$ be the unique polynomials in $\sqrt{X}$ such that the $X$-degree of $r(n, d, X)$ is strictly less than $1 + (nd/2)$ and

$$\mathfrak{G}(n, d, X) = q(n, d, X) \cdot X^{1 + \frac{nd}{2}} + r(n, d, X).$$

By the *support* of $q(n, d, X)$, we mean the set of half-integers $\varepsilon$ for which the coefficient of $X^\varepsilon$ in $q(n, d, X)$ is nonzero. Firstly, since $(2\ell - N + 1) \geq 0$, we infer that for $L \geq 0$,

$$g(N, \ell, L) \text{ is the coefficient of } X^L \text{ in } (-1) \cdot q(2\ell - N + 1, N, X).$$

Secondly, since the possible nonnegative values of the total angular momentum $L$ are the ones with $g(N, \ell, L) \neq 0$, they constitute the support of the polynomial (in $\sqrt{X}$) denoted by $q(2\ell - N + 1, N, X)$.

Let $N \geq 3$ be an integer and let $m$ be a positive integer not exceeding $1 + (N/2)$. Consider the Jain Mean Field picture with $m$ QEs above the $\nu = 1/3$ IQL state of the rest of the electrons as in Fig. 2.15.

As seen above, the set of possible (nonnegative) angular momenta for such a configuration is the support of $q(2\ell_1^* - m + 1, m, X)$; this support is denoted by

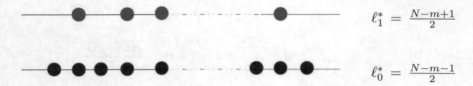

$$\ell_1^* = \frac{N-m+1}{2}$$

$$\ell_0^* = \frac{N-m-1}{2}$$

**Fig. 2.15**  Jain MFCF picture

$\Lambda(N, m)$. Note that $2\ell_1^* - m + 1 = N - 2m + 2$ and $(1/2)m(N - 2m + 2)$ is in $\Lambda(N, m)$, i.e., it is an allowed value of the total angular momentum $L$. Given an $L$ in $\Lambda(N, m)$, we want to construct a correlation function $G(z_1, \ldots, z_N)$ that is a nonzero homogeneous polynomial of total degree

$$\kappa_G := \frac{N(2(N-1)-m)}{2} - L$$

and such that the $z_i$-degree of $G(z_1, \ldots, z_N)$ is at most $2(N-1) - m$ for $1 \leq i \leq N$; for the sake of clarity, we prefer to denote our $G$ by $G_L$ (this is necessary especially when more than one value of $L$ is possible). In order for 0 to be in $\Lambda(N, m)$, the integer $Nm$ has to be even. More generally, from assertion (v) of Theorem 5, it follows that $L$ is in $\Lambda(N, m)$ only if there exists at least one nonzero semi-invariant of weight $m(N - 2m + 2)/2 - L$ and of degree at most $m$ (for the binary form of degree $N - 2m + 2$). For example, when $N = 11$ and $m = 3$, we have $N - 2m + 2 = 7$,

$$\mathfrak{G}(7, 3, X) = 1 + \sum_{i=2}^{9} X^i + X^6 - \sum_{i=13}^{22} X^i - X^{16},$$

$$\Lambda(11, 3) = \left\{ \tfrac{3}{2}, \tfrac{5}{2}, \tfrac{7}{2}, \tfrac{9}{2}, \tfrac{11}{2}, \tfrac{13}{2}, \tfrac{15}{2}, \tfrac{17}{2}, \tfrac{21}{2} \right\}.$$

If $L = 0$ is allowed, i.e., if $mN$ is even, then the corresponding correlation polynomial $G_0$ is necessarily a binary invariant of type $(N, 2(N-1) - m)$. In contrast, if $L > 0$, then $G_L$ is not a binary invariant of type $(N, d)$ for any $d$ (see Theorems 4 and 5); nevertheless, since $G_L$ is obtained by symmetrizing $\delta(z, E)$ for some $E \in E(N)$, it is indeed a semi-invariant of the binary form of degree $N$, i.e., a homogeneous, symmetric, translation-invariant polynomial in $z_1, \ldots, z_N$. In most of the constructions described below, where various correlation functions $G$ are realized as $Symm_N(\delta(z, E))$, the associated $E \in E(N, \leq 2(N-1) - m)$ has $2D_{N-m}$ as a diagonal-block and simultaneously $bound(E)$ is as small as possible (verification of this fact is straightforward; the details are left to the reader).

$\boxed{m = 1:}$

Suppose we have a single QE, i.e., $m = 1$. Then

$$\mathfrak{G}(N - 2m + 2, m, X) = \mathfrak{G}(N, 1, X) = (1 - X^{N+1}).$$

**Table 2.11** Correlation-statistics; 1 QE in $\nu = 1/3$ IQL

| correl. potency | 0 | 1 | 2 |
|---|---|---|---|
| No. of pairs | 1 | $N-2$ | $\frac{(N-1)(N-2)}{2}$ |
| Proportion | $\frac{2}{N(N-1)}$ | $\frac{2(N-2)}{N(N-1)}$ | $\frac{N-2}{N-1}$ |

Consequently, the only possible value of $L$ in this case is $N/2$. Let $G_{N/2} := Symm_N(\delta(z, E))$, where $E := E_{(1,N-1)}(N; 1, 1, 1)$ (see Corollary of Theorem 10), i.e., in block-form

$$E = \begin{bmatrix} 0 & u \\ u^T & 2D_{N-1} \end{bmatrix}, \quad \text{where } u := [0, 1, 1, \ldots, 1].$$

Note that $symgrp(E)$ is isomorphic to $S_{N-2}$. For notational simplicity, let $G := G_{N/2}$. Now assertion (iii) of the Corollary of Theorem 10 ensures that $G(z_1, \ldots, z_N)$ is a nonzero homogeneous polynomial of total degree

$$\kappa_G := \frac{N(2N-3)}{2} - \frac{N}{2} = N(N-2),$$

and its $z_i$-degree is at most $2N - 3$ for $1 \le i \le N$. Clearly,

$$suppt(E) = \{0, 1, 2\}, \qquad bound(E) = 2$$

and

$$frq(b, E) = \begin{cases} 1 & \text{if } b = 0, \\ N-2 & \text{if } b = 1, \\ \frac{(N-1)(N-2)}{2} & \text{if } b = 2. \end{cases}$$

So, we have the correlation-statistics as it appears in Table 2.11.

Note that, as expected, when $N \to \infty$, the proportion of 2-correlated Fermion-pairs tends to 1.

$\boxed{m = 2 :}$

Consider the case of two QEs, i.e., $m = 2$. In this case,

$$\mathfrak{G}(N - 2m + 2, m, X) = \mathfrak{G}(N - 2, 2, X) = \frac{(1 - X^{N-1})(1 - X^N)}{(1 - X^2)}.$$

Now it is straightforward to verify that

$$\Lambda(N, m) = \left\{ p(N) + 2r \,\middle|\, r \in \mathbb{N}, \quad 0 \le r \le \frac{1}{2}(N - 2 - p(N)) \right\},$$

where $p(N)$ is the *parity of* $N$, i.e.,

$$p(N) := \begin{cases} 0 \text{ if } N \text{ is even,} \\ 1 \text{ if } N \text{ is odd.} \end{cases}$$

For $1 \leq i \leq N - 2$, define $R_i := [2, \ 0]$ if $i$ is odd and $R_i := [0, \ 2]$ if $i$ is even. Let $A$ be the $(N - 2) \times 2$ matrix having $R_i$ as its $i$th row. For $0 \leq r \leq (1/2)(N - 2 - p(N))$, let $E_r \in E(N)$ be the matrix defined in block-form by

$$E_r := \begin{bmatrix} a_r D_2 & A^T \\ \\ A & 2D_{N-2} \end{bmatrix}, \quad \text{where } a_r := N - 2 - p(N) - 2r.$$

Assertion (i) of Theorem 7 ensures that for $0 \leq r \leq (1/2)(N - 2 - p(N))$,

$$\text{letting } L = p(N) + 2r \quad \text{and} \quad G_L := Symm_N(\delta(z, E_r)),$$

$G_L$ is a nonzero polynomial which is homogeneous of total degree

$$\kappa_{G_L} := \frac{N(2N - 4)}{2} - L = N(N - 2) - p(N) - 2r,$$

and its $z_i$-degree is at most $2N - 4$ for $1 \leq i \leq N$. Clearly,

$$suppt(E_r) = \{0, \ 2, \ N - 2 - p(N) - 2r\}, \quad bound(E_r) = \max\{2, \ N - 2 - p(N) - 2r\}.$$

If $L < N - 4$, then

$$frq(b, E_r) = \begin{cases} N - 2 & \text{if } b = 0, \\ \frac{(N-1)(N-2)}{2} & \text{if } b = 2, \\ 1 & \text{if } b = N - 2 - p(N) - 2r. \end{cases}$$

If $L = N - 2$, then

$$frq(0, E_r) = N - 1 \quad \text{and} \quad frq(2, E_r) = \frac{(N - 1)(N - 2)}{2}.$$

If $L = N - 4$, then

$$frq(0, E_r) = N - 2 \quad \text{and} \quad frq(2, E_r) = 1 + \frac{(N - 1)(N - 2)}{2}.$$

For $L = N - 2$, the resulting correlation-statistics is presented in Table 2.12.

Likewise, if $L \leq N - 4$, then the correlation-statistics is tabulated in Table 2.13.

Of course, here too, as $N \to \infty$ the proportion of 2-correlated Fermion-pairs tends to 1.

**Table 2.12** Correlation-statistics; 2 QE in $\nu = 1/3$ IQL ($L = N - 2$)

| correl. potency | 0 | 2 |
|---|---|---|
| No. of pairs | $N - 1$ | $\frac{(N-1)(N-2)}{2}$ |
| Proportion | $\frac{2}{N}$ | $\frac{N-2}{N}$ |

**Table 2.13** Correlation-statistics; 2 QE in $\nu = 1/3$ IQL ($L \leq N - 4$)

| $L$ | $<N - 4$ | | | $N - 4$ | |
|---|---|---|---|---|---|
| correl. potency | 0 | 2 | $N - L - 2$ | 0 | 2 |
| No. of pairs | $N - 2$ | $\frac{(N-1)(N-2)}{2}$ | 1 | $N - 2$ | $\frac{N^2-3N+4}{2}$ |
| Proportion | $\frac{2(N-2)}{N(N-1)}$ | $\frac{N-2}{N}$ | $\frac{2}{N(N-1)}$ | $\frac{2(N-2)}{N(N-1)}$ | $\frac{N^2-3N+4}{N(N-1)}$ |

$$\boxed{m = \frac{1}{2}N :}$$

Consider the case where $N \geq 4$ is even and $m = N/2$. Then

$$\mathfrak{G}(N - 2m + 2, m, X) = \mathfrak{G}(2, m, X) = \frac{(1 - X^{m+1})(1 - X^{m+2})}{(1 - X^2)}.$$

It is straightforward to verify that

$$\Lambda(N, m) = \left\{ m - 2r \mid r \in \mathbb{N}, \ \ 0 \leq r \leq \frac{m}{2} \right\}.$$

For $0 \leq r \leq m/2$, let $A_r$ be the $m \times m$ symmetric matrix $[a_{ij}]$, where

$$a_{ij} := \begin{cases} 2 \text{ if } i = j, \\ 0 \text{ if } \{i, j\} = \{2s - 1, 2s\} \text{ for some } 0 \leq s \leq r, \\ 1 \text{ otherwise.} \end{cases}$$

We observe that for each $r$, $A_r$ satisfies the requirements imposed on the matrix $A$ of Theorem 11. Clearly, our $N$ and $m$ satisfy the hypotheses of Theorem 11 and $A_r$ is indeed an $m \times (N - m)$ symmetric matrix. Finally, letting $a := 1$, we have $a_{ii} = 2 \geq 2a$ for $1 \leq i \leq m$ as well as $2a > a_{ij}$ for $1 \leq i < j \leq m$. Define $E_r \in E(N)$ by setting

$$E_r := 2D_N - B_r, \quad \text{where} \quad B_r := \begin{bmatrix} 0 & A_r \\ A_r & 0 \end{bmatrix},$$

**Table 2.14**  Correlation-statistics; $N/2$ QE in $\nu = 1/3$ IQL with $L = (N - 4r)/2$

| correl. potency | 0 | 1 | 2 |
|---|---|---|---|
| No. of pairs | $\frac{N}{2}$ | $\frac{N^2-2N-8r}{4}$ | $\frac{N^2-2N+8r}{4}$ |
| Proportion | $\frac{1}{N-1}$ | $\frac{N^2-2N-8r}{2N(N-1)}$ | $\frac{N^2-2N+8r}{2N(N-1)}$ |

and for each $L(r) := m - 2r$, let $G_{L(r)} := Symm_N(\delta(z, E_r))$. Then, the polynomial $G_{L(r)}$ is homogeneous in $z_1, \ldots, z_N$ of total degree

$$\kappa_{G_{L(r)}} := \frac{N(3N - 4)}{4} - L(r) = m(3m - 2) - m + 2r,$$

and its $z_i$-degree is at most $2(N - 1) - m = 3m - 2$ for $1 \leq i \leq N$. Most importantly, by Theorem 11, we have $Symm_N(\delta(z, -B_r)) \neq 0$ for $0 \leq r \leq m/2$. Now since

$$G_{L(r)} = Symm_N(\delta(z, E_r)) = Symm_N\left(\frac{\delta(z, 2D_N)}{\delta(z, B_r)}\right)$$

and equally plainly, we have

$$Symm_N\left(\frac{\delta(z, 2D_N)}{\delta(z, B_r)}\right) = \delta(z, 2D_N) \cdot Symm_N(\delta(z, -B_r)) \neq 0,$$

it follows that $G_{L(r)} \neq 0$. In passing, we note that $E_0 = M(m, 2, 1, 1)$. Clearly

$$suppt(E_r) = \{0, 1, 2\}, \quad bound(E_r) = 2.$$

Also, for $0 \leq r \leq m/2$,

$$frq(b, E_r) = \begin{cases} \frac{N}{2} & \text{if } b = 0, \\ \frac{N^2-2N-8r}{4} & \text{if } b = 1, \\ \frac{N^2-2N+8r}{4} & \text{if } b = 2. \end{cases}$$

So, for each integer $r$ with $0 \leq r \leq m/2$, Table 2.14 exhibits the corresponding correlation-statistics.

As $N \to \infty$, the proportion of 1-correlated and 2-correlated pairs each tends to $1/2$. This is in accordance with the fact that half of the particles are quasi-electrons.

Consider the special case where $m$ is an even integer (whence $N/4$ is an integer). As assured by (i) of the Lemma preceding Theorem 12,

$$\mathbb{M}(m, m, m/2, r + m(m - 1)/2) \neq \emptyset \quad \text{for} \quad 0 \leq r \leq m/2.$$

So, for $0 \leq r \leq m/2$, pick a $C_r \in \mathbb{M}(m, m, m/2, r + m(m - 1)/2)$ and define

$$E_r := \begin{bmatrix} 2D_m & 2C_r \\ \\ 2C_r^T & 2D_m \end{bmatrix}.$$

Assertion (i) of Theorem 7 ensures that for $L = m - 2r$, the polynomial $G_L := Symm_N(\delta(z, E_r))$ is a nonzero homogeneous polynomial of total degree $m(3m - 2) - m + 2r$ and its $z_i$-degree is at most $2(N - 1) - m = 3m - 2$ for $1 \le i \le N$. Also, $suppt(E_r) = \{0, 2\}$ and $bound(E_r) = 2$. More concretely, let

$$C_{m/2} := \mathrm{cirmat}((a_1, \ldots, a_m)),$$

where $a_i = 1$ for $1 \le i \le m/2$ and $a_i = 0$ otherwise. Then, for $0 \le r \le m/2$, let $C_r$ be obtained from $C_{m/2}$ by replacing any (randomly picked) $(m/2) - r$ entries 1 in $C_{m/2}$ by 0. For example, here is a list of possible $2C_r$ when $N = 8$ and $m = 4$.

$$2C_2 := \begin{bmatrix} 2 & 2 & 0 & 0 \\ 0 & 2 & 2 & 0 \\ 0 & 0 & 2 & 2 \\ 2 & 0 & 0 & 2 \end{bmatrix}, \quad 2C_1 := \begin{bmatrix} 2 & 2 & 0 & 0 \\ 0 & 2 & 2 & 0 \\ 0 & 0 & 2 & 2 \\ 0 & 0 & 0 & 2 \end{bmatrix},$$

$$2C_0 := \begin{bmatrix} 0 & 2 & 0 & 0 \\ 2 & 0 & 2 & 0 \\ 0 & 2 & 0 & 2 \\ 0 & 0 & 2 & 0 \end{bmatrix}.$$

In comparison to the configurations described in the first part, these special configurations have a higher proportion of correlation potency 2 factors and hence represent higher energy states. Tabulation of the correlation-statistics for these special configurations is left to the reader.

$$\boxed{m = \frac{1}{2}(N + 1):}$$

Consider the case where $N \ge 5$ is odd and $m = (N + 1)/2$. Then

$$\mathfrak{G}(N - 2m + 2, m, X) = \mathfrak{G}(1, m, X) = (1 - X^{m+1}).$$

Letting $N := 2n + 1$, we have $m = n + 1$ and

$$\Lambda(N, m) = \left\{ \frac{N + 1}{4} \right\} = \left\{ \frac{n + 1}{2} \right\}.$$

Let $A$ be the $n \times (n + 1)$ matrix $[a_{ij}]$ such that for $1 \le i \le n$ and $1 \le j \le n + 1$,

$$a_{ij} := \begin{cases} 1 & \text{if } i \ne j \text{ and } (i, j) \ne (n, n + 1), \\ 2 & \text{otherwise.} \end{cases}$$

Let $E \in E(N)$ be the matrix defined in block-form by

$$
E := 2D_N - \begin{bmatrix} 0 & A \\ A^T & 0 \end{bmatrix}.
$$

Recalling the definitions just preceding Theorem 10, it is easily verified that

$$
grp(A) = \{\theta \in S_n \mid \theta(n) = n\}
$$

and $rat(A, T) = \{pol(A, T)^{-1}\}$, where

$$
pol(A, T) = \prod_{r=1}^{n-1}(T_r - T_n) \prod_{1 \le r < s \le n} (T_r - T_s)^2.
$$

As a consequence, $A$ is seen to be an admissible matrix. Now Theorem 10 allows us to conclude that $G := Symm_N(\delta(z, E))$ is a nonzero polynomial which is homogeneous of total degree

$$
\kappa_G := \frac{N(3N - 5)}{4} - \frac{(N + 1)}{4} = N(N - 1) - \frac{(N + 1)^2}{4},
$$

and its $z_i$-degree is at most $2(N - 1) - m = 3n - 1$ for $1 \le i \le N$. Clearly,

$$
suppt(E) = \{0, 1, 2\}, \quad bound(E) = 2.
$$

Likewise, we note that

$$
frq(b, E) = \begin{cases} \frac{N+1}{2} & \text{if } b = 0, \\ \frac{(N-3)(N+1)}{4} & \text{if } b = 1, \\ \frac{(N-1)^2}{4} & \text{if } b = 2. \end{cases}
$$

The corresponding correlation-statistics is presented in Table 2.15.

$$
\boxed{m = 1 + \frac{1}{2}N :}
$$

**Table 2.15** Correlation-statistics; $(N + 1)/2$ QE in $\nu = 1/3$ IQL

| correl. potency | 0 | 1 | 2 |
|---|---|---|---|
| No. of pairs | $\frac{N+1}{2}$ | $\frac{(N-3)(N+1)}{4}$ | $\frac{(N-1)^2}{4}$ |
| Proportion | $\frac{N+1}{N(N-1)}$ | $\frac{(N-3)(N+1)}{2N(N-1)}$ | $\frac{N-1}{2N}$ |

Consider the case where $N$ is even and $m = 1 + (N/2)$. Observe that $\Lambda(N, 1 + (N/2)) = \{0\}$ and hence $L = 0$. Let $G := Symm_N(\delta(z, E))$, where $E := M(N/2, 2, 1, 1)$. For even $m$, we also have the option of letting $G := Symm_N(\delta(z, E_0))$, where $E_0 := M_0(N/2, 2, 1, 1)$. As an aside, we remark that this construction of $G$ is identical to the one employed above for the IQL state of $N$ fermions corresponding to $\nu = 2/5$. In particular, the correlation-statistics is also that of the IQL state of $N$ fermions corresponding to $\nu = 2/5$. If $N = 4$, then since the space of binary invariants of type $(4, 3)$ has dimension 1, our $G$ is essentially (i.e., up to numerical multiples) the only nonzero binary invariant of type $(4, 3)$.

### Other systems :

For arbitrary values of $N$ and $m$, no explicit description of the set $\Lambda(N, m)$ is known. If $N$ is an even integer $\geq 7$ and $3 \leq m < N/2$, then $L = 0$ is indeed a possible value of the total angular momentum; but, even in this case, we are unable to provide an existent configuration. At present, we remain content with a full and detailed consideration of the cases with $N \leq 8$. With this restriction on $N$, only two systems remain to be dealt with; all others are special cases of what has been established above. Namely, we have to consider the system $(N, m) = (7, 3)$ and the system $(N, m) = (8, 3)$. To address these two, we need to recall that $D_{r,s}$ denotes the $r \times s$ matrix whose $ij$th entry is $(1 - \delta_{ij})$, where $\delta_{ij}$ is the *Kronecker delta*, and $D_r = D_{r,r}$.

### $(N = 7, m = 3)$ :

In this case, $2(N - 1) - m = 9$ and

$$\Lambda(7, 3) = \left\{ \frac{3}{2}, \frac{5}{2}, \frac{9}{2} \right\}.$$

For $L \in \Lambda(7, 3)$, let $G_L := Symm_N(\delta(z, A_L))$, where $A_L \in E(7)$ is defined as follows:

$$A_{3/2} := 2D_7 - \begin{bmatrix} 0 & C_{3/2} \\ C_{3/2}^T & 0 \end{bmatrix}, \quad \text{where } C_{3/2} := \begin{bmatrix} 2 & 0 & 1 & 1 \\ 1 & 1 & 0 & 2 \\ 0 & 2 & 2 & 0 \end{bmatrix},$$

$$A_{5/2} := 2D_7 - \begin{bmatrix} 0 & C_{5/2} \\ C_{5/2}^T & 0 \end{bmatrix}, \quad \text{where } C_{5/2} := \begin{bmatrix} 2 & 1 & 1 & 1 \\ 1 & 1 & 0 & 2 \\ 0 & 2 & 2 & 0 \end{bmatrix},$$

$$A_{9/2} := 2D_7 - \begin{bmatrix} 0 & C_{9/2} \\ C_{9/2}^T & 0 \end{bmatrix}, \quad \text{where } C_{9/2} := \begin{bmatrix} 2 & 1 & 1 & 1 \\ 1 & 1 & 1 & 2 \\ 0 & 2 & 2 & 1 \end{bmatrix}.$$

Of course, $Symm_7(\delta(z, 2D_7 - E)) = \delta(z, 2D_7) \cdot Symm_7(\delta(z, -E))$ for any $E \in E(7)$. So, it suffices to show that $Symm_7(\delta(z, A_s - 2D_7)) \neq 0$ for $s = 3/2, 5/2, 9/2$ and hence, in view of Theorem 10, it suffices to show that $C_s$ is admissible for $s =$

$3/2, 5/2, 9/2$. It is straightforward to verify that $grp(C_s) = \{id, (1, 2)\} < S_3$ for $s = 3/2, 5/2, 9/2$. Also, $rat(C_{3/2}, T)$, $rat(C_{5/2}, T)$ and $rat(C_{9/2}, T)$ are

$$\{(T_1 - T_2)^{-2}(T_1 - T_3)^{-1}(T_2 - T_3)^{-1}\},$$
$$\{(T_1 - T_2)^{-2}(T_1 - T_3)^{-2}(T_2 - T_3)^{-1}, \ (T_1 - T_2)^{-2}(T_2 - T_3)^{-2}(T_1 - T_3)^{-1}\},$$
$$\{(T_1 - T_2)^{-2}(T_1 - T_3)^{-2}(T_2 - T_3)^{-3}, \ (T_1 - T_2)^{-2}(T_2 - T_3)^{-2}(T_1 - T_3)^{-3}\},$$

respectively. Clearly $rat(C_{3/2}, T)$ is $\mathbb{Q}$-linearly independent. Since $\{(T_1 - T_3), (T_2 - T_3)\}$ is $\mathbb{Q}$-linearly independent, each of $rat(C_{5/2}, T)$, $rat(C_{9/2}, T)$ is $\mathbb{Q}$-linearly independent. Thus, $G_{3/2}$, $G_{5/2}$ and $G_{9/2}$ are nonzero homogeneous polynomials of total degrees 30, 29 and 27, respectively. Moreover, the $z_i$-degree of each $G_L$ is at most 9 for $1 \leq i \leq 7$.

$\boxed{(N = 8, \ m = 3):}$

In this case, $2(N - 1) - m = 11$ and

$$\Lambda(8, 3) = \{0, 2, 3, 4, 6\}.$$

For $L \in \Lambda(8, 3)$, let $G_L := Symm_N(\delta(z, A_L))$, where $A_L \in E(8)$ is defined as follows:

$$A_0 := \begin{bmatrix} 3D_3 & C \\ C^T & 2D_5 \end{bmatrix}, \quad \text{where} \quad C := \begin{bmatrix} 2 & 0 & 1 & 2 & 0 \\ 1 & 2 & 0 & 1 & 1 \\ 0 & 1 & 2 & 0 & 2 \end{bmatrix},$$

$$A_2 := \begin{bmatrix} 3D_3 & C \\ C^T & 2D_5 \end{bmatrix}, \quad \text{where} \quad C := \begin{bmatrix} 2 & 0 & 1 & 2 & 0 \\ 1 & 2 & 0 & 0 & 0 \\ 0 & 1 & 2 & 0 & 2 \end{bmatrix},$$

$$A_3 := \begin{bmatrix} 2D_3 & C \\ C^T & 2D_5 \end{bmatrix}, \quad \text{where} \quad C := \begin{bmatrix} 2 & 0 & 1 & 2 & 0 \\ 0 & 1 & 2 & 0 & 1 \\ 1 & 2 & 0 & 1 & 2 \end{bmatrix},$$

$$A_4 := \begin{bmatrix} 2D_3 & C \\ C^T & 2D_5 \end{bmatrix}, \quad \text{where} \quad C := \begin{bmatrix} 0 & 0 & 1 & 2 & 1 \\ 2 & 1 & 0 & 1 & 1 \\ 1 & 2 & 1 & 0 & 1 \end{bmatrix},$$

$$A_6 := \begin{bmatrix} 2D_3 & D_{3,5} \\ D_{5,3} & 2D_5 \end{bmatrix}, \quad \text{i.e.,} \quad M(0 < 3 < 8, 1, 1).$$

Observe that for each $L \in \{0, 2, 3, 4, 6\}$, $G_L$ is homogeneous of total degree $44 - L$ and its $z_i$-degree does not exceed 11 for $1 \leq i \leq 8$. Assertion (ii) of Theorem 8 is directly applicable to the $L = 6$ case, ensuring that $G_6 \neq 0$. Nontriviality of $G_0, G_2$ does not follow from any of our theorems. Nevertheless, a SAGE computation shows that the evaluation of each of $G_0, G_2$ at $z_i = i - 1$ for $1 \leq i \leq 8$ is a nonzero integer and hence, apriori, $G_0$, $G_2$ are nonzero. Lastly, we show how Theorem 10 can be used to verify that $G_3$ and $G_4$ are nonzero. To begin with, note that

$$A_3 := 2D_8 - \begin{bmatrix} 0 & A \\ A^T & 0 \end{bmatrix}, \quad \text{where} \quad A := \begin{bmatrix} 0 & 2 & 1 & 0 & 2 \\ 2 & 1 & 0 & 2 & 1 \\ 1 & 0 & 2 & 1 & 0 \end{bmatrix},$$

and

$$A_4 := 2D_8 - \begin{bmatrix} 0 & B \\ B^T & 0 \end{bmatrix}, \quad \text{where} \quad B := \begin{bmatrix} 2 & 2 & 1 & 0 & 1 \\ 0 & 1 & 2 & 1 & 1 \\ 1 & 0 & 1 & 2 & 1 \end{bmatrix}.$$

Again, $Symm_8(\delta(z, 2D_8 - E)) = \delta(z, 2D_8) \cdot Symm_8(\delta(z, -E)$ for any $E \in E(8)$. So, it suffices to show that $Symm_8(\delta(z, A_s - 2D_8)) \neq 0$ for $s = 3, 4$. In the very first example of the set of examples preceding Theorem 10, it is verified that $A$ is an admissible matrix with $max(A) = 2$. Hence by Theorem 10, $Symm_8(\delta(z, A_3 - 2D_8)) \neq 0$. In the case of $B$, it is straightforward to verify that $grp(B) = \{id, (2, 3)\} < S_3$ and $rat(B, T) = \{\rho_1, \rho_2\}$, where

$$\rho_1 = T_1^{-1} T_2^{-1} T_3^{-1} (T_1 - T_2)^{-2} (T_1 - T_3)^{-1} (T_2 - T_3)^{-2},$$

$$\rho_2 = T_1^{-1} T_2^{-1} T_3^{-1} (T_1 - T_3)^{-2} (T_1 - T_2)^{-1} (T_2 - T_3)^{-2}.$$

Since the $T_3$-degrees of the rational functions in $rat(B, T)$ are clearly distinct, $rat(B, T)$ is $\mathbb{Q}$-linearly independent. It readily follows that $B$ is admissible and $max(B) = 2$. Thanks to Theorem 10, $Symm_8(\delta(z, A_4 - 2D_8)) \neq 0$.

*Remarks*

1. It is easy to see that for any choice of $A_0$, $A_2$ in the above $(N, m) = (8, 3)$ case, at least one entry in each of $A_0$, $A_2$ has to be $\geq 3$.
2. In the case of $(N, m) = (8, 3)$, disregarding the requirement of $2D_5$ as a diagonal block leads to more choices. Below, we provide a sample of some (but certainly not all) alternative choices of $A_L$; these also serve to demonstrate the utility of our theorems.

$$A_0 := M(2, 4, 1, 3).$$

Then assertion (ii) of the Corollary of Theorem 8 ensures that $G_0$ is nonzero.

$$A_2 := 2D_8 - \begin{bmatrix} 0 & C \\ C & 0 \end{bmatrix}, \quad \text{where} \quad C := \begin{bmatrix} 2 & 1 & 0 & 0 \\ 1 & 2 & 1 & 0 \\ 0 & 1 & 2 & 1 \\ 0 & 0 & 1 & 2 \end{bmatrix}.$$

Then Theorem 10 ensures that $G_2$ is nonzero. Yet another choice for $A_2$ is

$$A_2 := \begin{bmatrix} 0 & C \\ C & 0 \end{bmatrix}, \quad \text{where} \quad C := \begin{bmatrix} 3 & 2 & 2 & 2 \\ 2 & 3 & 3 & 3 \\ 2 & 3 & 3 & 3 \\ 2 & 3 & 3 & 3 \end{bmatrix}.$$

Then assertion (i) of Theorem 8 ensures that $G_2$ is nonzero.

$$A_3 := \begin{bmatrix} 0 & u & v \\ u^T & 0 & C \\ v^T & C^T & 0 \end{bmatrix},$$

where $u := [1, 1, 1]$, $v := [2\,2\,2\,2]$ and

$$C := \begin{bmatrix} 1 & 3 & 3 & 3 \\ 1 & 3 & 3 & 3 \\ 1 & 3 & 3 & 3 \end{bmatrix}.$$

Then (i) of Theorem 8 allows us to infer that the corresponding $G_3$ is nonzero.

$$A_4 := \begin{bmatrix} B & C \\ C^T & 2D_5 \end{bmatrix}, \quad \text{where} \quad B := \begin{bmatrix} 0 & 2 & 4 \\ 2 & 0 & 4 \\ 4 & 4 & 0 \end{bmatrix} \quad \text{and}$$

$$C := \begin{bmatrix} 2 & 0 & 0 & 2 & 0 \\ 0 & 2 & 0 & 0 & 2 \\ 0 & 0 & 2 & 0 & 0 \end{bmatrix}.$$

Then assertion (i) of Theorem 7 ensures that $G_4$ is nonzero.

$$A_6 := \begin{bmatrix} 0 & C \\ C^T & 0 \end{bmatrix}, \quad \text{where} \quad C := \begin{bmatrix} 2 & 2 & 2 & 4 \\ 2 & 2 & 2 & 4 \\ 2 & 2 & 2 & 2 \\ 4 & 4 & 2 & 0 \end{bmatrix}.$$

Then assertion (i) of Theorem 7 implies that $G_6$ is nonzero.

## Systems of $N \leq 8$ Fermions with QE in the $\nu = 1/3$ IQL

As a demonstration of the results established above, we proceed to exhibit the correlation diagrams of configurations with quasielectrons for $4 \leq N \leq 8$. In each diagram, quasielectrons are represented by the red vertices. The correlation function corresponding to any one of these configurations, when presented as a polynomial in $z_1, \ldots, z_N$, has a very large number of terms appearing in it; so, it is not very useful

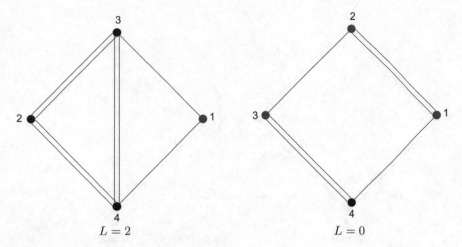

**Fig. 2.16**  $N = 4$; 1 QE and 3 QE in the $\nu = 1/3$ IQL

to explicitly exhibit these polynomials. They are presented only in the case of $N = 4$, since in that case the associated correlation polynomials have compact expressions as polynomials in $y_1, y_2, y_3$, where (as in the definitions preceding Theorem 1)

$$y_1 = -3/8\,z_1^2 + 1/8\,(2\,z_2 + 2\,z_3 + 2\,z_4)\,z_1 - 3/8\,z_2^2$$
$$+ 1/8\,(2\,z_3 + 2\,z_4)\,z_2 - 3/8\,z_3^2 + 1/4\,z_3 z_4 - 3/8\,z_4^2,$$
$$y_2 = 1/8\,(-z_4 + z_1 - z_3 + z_2)\,(z_4 + z_1 - z_3 - z_2)\,(-z_4 + z_1 + z_3 - z_2),$$

and $y_3 = (-1/256) \cdot P_1 \cdot P_2$, where

$$P_1 = (-3\,z_4 + z_1 + z_3 + z_2)\,(z_4 + z_1 - 3\,z_3 + z_2) \quad \text{and}$$
$$P_2 = (3\,z_1 - z_2 - z_3 - z_4)\,(-3\,z_2 + z_3 + z_4 + z_1).$$

If $N \geq 5$, expressions of the associated correlation functions as polynomials in $y_1, \ldots, y_{N-1}$ are also too long and complicated to write down explicitly.

Consider configurations with $N = 4$ and either 1 QE or 3 QE (as in Fig. 2.16). Then, the respective correlation functions are $G_2$ and $G_0$, where the suffixes 2, 0 stand for the total angular momentum of the system. Then, we have

$$G_2 = -16\,y_1^4 + 256\,y_3 y_1^2 - 144\,y_1 y_2^2 - 768\,y_3^2,$$
$$G_0 = -8\,y_1^3 + 288\,y_3 y_1 - 108\,y_2^2.$$

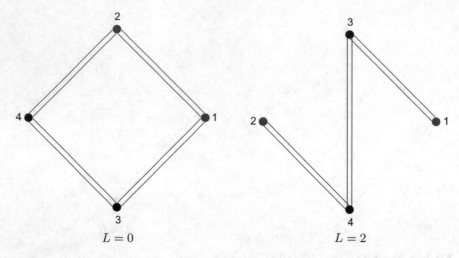

**Fig. 2.17** $N = 4$; 2 QE in the $\nu = 1/3$ IQL

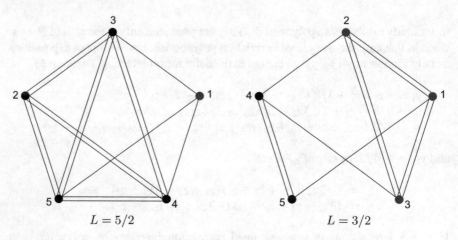

**Fig. 2.18** $N = 5$; 1 QE and 3 QE in the $\nu = 1/3$ IQL

Next, consider the configurations with $N = 4$ and 2 QE (as in Fig. 2.17). Here, the possible values of the total angular momentum are $L = 0, 2$ and the respective correlation functions $G_0$ and $G_2$ are given by

$$G_0 = 8\left(y_1{}^2 + 12\,y_3\right)^2 \quad \text{and}$$
$$G_2 = -24\,y_1{}^3 - 160\,y_3 y_1 - 36\,y_2{}^2.$$

The rest of this section exhibits the existent configurations of $5 \leq N \leq 8$ systems with QE in the $\nu = 1/3$ IQL (Figs. 2.18, 2.19, 2.20, 2.21, 2.22, 2.23, 2.24, 2.25, 2.26, 2.27, 2.28, 2.29, 2.30, 2.31, 2.32, 2.33 and 2.34).

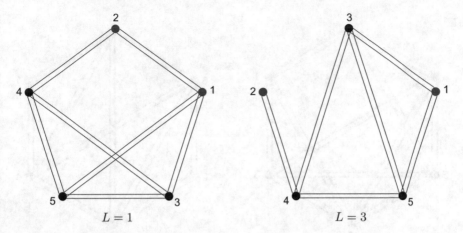

**Fig. 2.19** $N = 5$; 2 QE in the $\nu = 1/3$ IQL

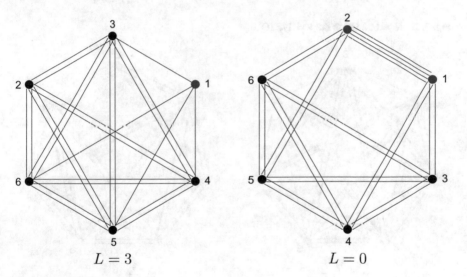

**Fig. 2.20** $N = 6$; 1 QE and 2 QE in the $\nu = 1/3$ IQL

In this special case of $N = 8$ and $m = 4$, Figs. 2.32 and 2.33 show the three alternative configurations mentioned earlier in dealing with the case $m = N/2$ and $m$ even.

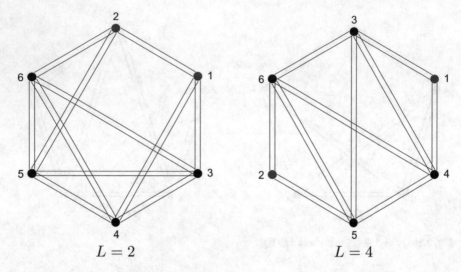

**Fig. 2.21** $N = 6$; 2 QE in the $\nu = 1/3$ IQL

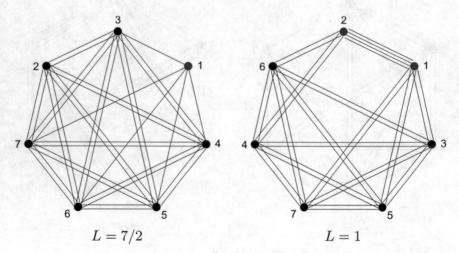

**Fig. 2.22** $N = 7$; 1 QE and 2 QE in the $\nu = 1/3$ IQL

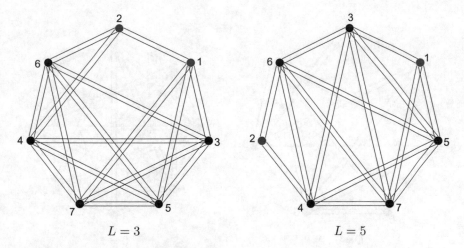

**Fig. 2.23**  $N = 7$; 2 QE in the $\nu = 1/3$ IQL

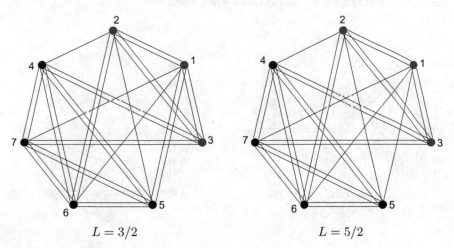

**Fig. 2.24**  $N = 7$; 3 QE in the $\nu = 1/3$ IQL

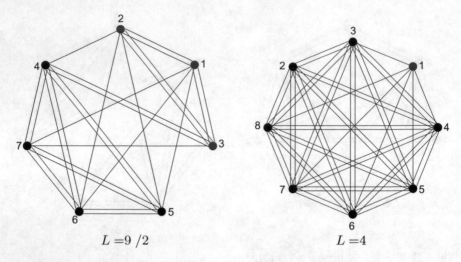

Fig. 2.25  $N = 7$; 3 QE and $N = 8$; 1 QE in the $\nu = 1/3$ IQL

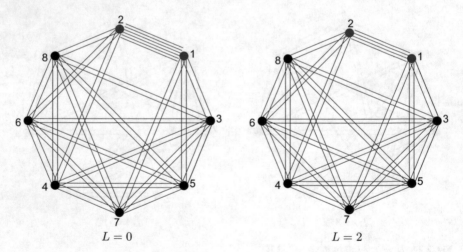

Fig. 2.26  $N = 8$; 2 QE in the $\nu = 1/3$ IQL

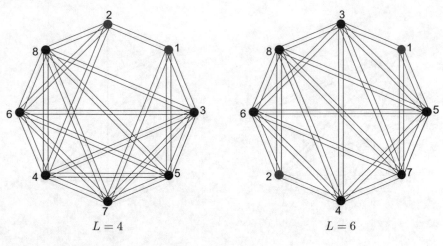

**Fig. 2.27** $N = 8$; 2 QE in the $\nu = 1/3$ IQL

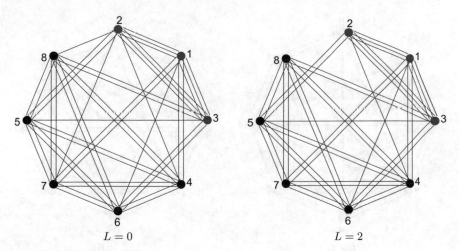

**Fig. 2.28** $N = 8$; 3 QE in the $\nu = 1/3$ IQL

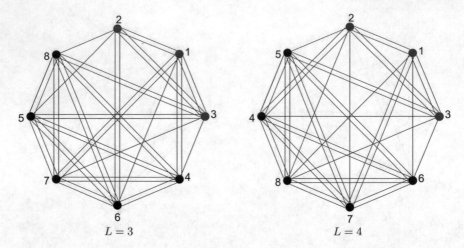

**Fig. 2.29** $N = 8$; 3 QE in the $\nu = 1/3$ IQL

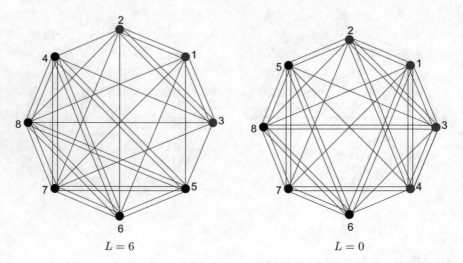

**Fig. 2.30** $N = 8$; 3 QE and 4 QE in the $\nu = 1/3$ IQL

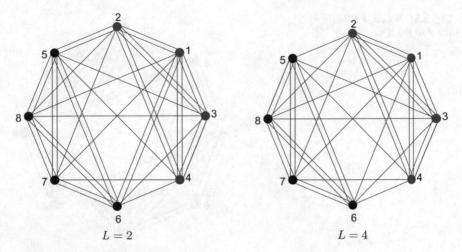

**Fig. 2.31** $N = 8$; 4 QE in the $\nu = 1/3$ IQL

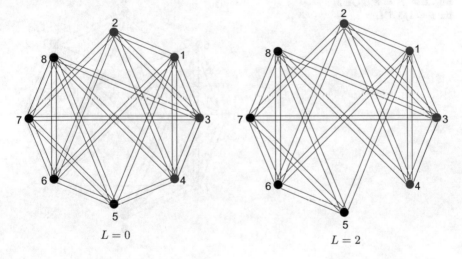

**Fig. 2.32** $N = 8$; 4 QE in the $\nu = 1/3$ IQL

**Fig. 2.33**  $N = 8$; 4 QE in
the $\nu = 1/3$ IQL

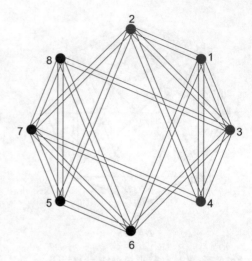

**Fig. 2.34**  $N = 8$; 5 QE in
the $\nu = 1/3$ IQL

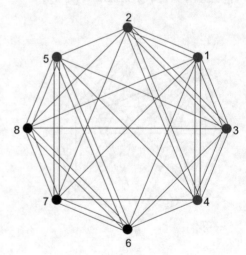

# References

1. C. Greenhill, B.D. McKay, Asymptotic enumeration of sparse multigraphs with given degrees. SIAM J. Discrete Math. **27**, 2064–2089 (2013)
2. B.D. McKay, A. Piperno, Practical graph isomorphism II. J. Symb. Comput. **60**, 94–112 (2014)
3. G.A. Baker, S.B. Mulay, Geometric and analytical aspects of anyons. Int. J. Theor. Phys. **34**, 2435–2451 (1995)
4. J.H. Grace, A. Young, *The Algebra of Invariants* (Chelsea Publishing Company, New York, 1964) (1903), reprint
5. J.P.S. Kung, G.-C. Rota, The invariant theory of binary forms. Bull. Am. Math. Soc. **10**, 27–85 (1984)

6. O. Zariski, P. Samuel, *Commutative Algebra*. Graduate Texts in Mathematics, vols. I and II (Springer, New York, 1976)
7. E.B. Elliot, *An Introduction to the Algebra of Quantics* (Chelsea Publishing Company, New York, 1964), 2nd edn. (1913), reprint
8. J. Dixmier, Quelques résultats et conjectures concernant les séries de Poincaré des invariants des formes binaires. *Séminaire d'algèbre Paul Dubreil et Marie-Paule Malliavin, 36ème année (Paris, 1983–1984)*. Lecture Notes in Mathematics, vol. 1146 (Springer, Berlin, 1985), pp. 127–160
9. A.E. Brouwer, J. Draisma, M. Popoviciu, The degrees of a system of parameters of the ring of invariants of a binary form, arXiv:1404.5722 (2014)
10. S.B. Mulay, J.J. Quinn, M.A. Shattuck, An algebraic approach to FQHE variational wave functions. arXiv:1808.10284 (2018)
11. S.B. Mulay, Graph-monomials and invariants of binary forms. arXiv:1809.00369 (2018)
12. S.B. Mulay, J.J. Quinn, M.A. Shattuck, An algebraic approach to electron interactions in quantum Hall systems. arXiv:1809.01504 (2018)
13. A. Abdesselam, J. Chipalkatti, Brill-Gordon loci, transvectants and an analogue of the Foulkes conjecture. Adv. Math. **208**, 491–520 (2007)
14. G. Sabidussi, Binary invariants and orientations of graphs. Discrete Math. **101**, 251–277 (1992)
15. J. Chipalkatti, On Hermite's invariant for binary quintics. J. Algebra **317**, 324–353 (2007)
16. R.F. Curl, J.E. Kilpatrick, Atomic term symbols by group theory. Am. J. Phys. **28**, 357–365 (1960)
17. A.T. Benjamin, J.J. Quinn, J.J. Quinn, A. Wójs, Composite Fermions and integer partitions. J. Comb. Theory Ser. A **95**, 390–397 (2001)

# Appendix A
# Moore-Read State

The trial wave function for $N = 2n$ Fermions in the Moore-Read state with $\nu = 2 + 1/2$, is denoted by $\Psi_{MR}$. It is well known that $\Psi_{MR}$ is a product of $\prod_{1 \leq i < j \leq N}(z_i - z_j)$ (the so called Fermi-factor) and a certain correlation polynomial $G_{MR}(z_1, \ldots, z_{2n})$ which is a product of the Fermi-factor with the "Pfaffian" $Pf(z_{ij}^{-1})$ defined below. As mentioned in the seventh section of the first chapter, an equivalent correlation polynomial can be obtained following the more intuitive approach of correlation diagrams (as in the previous sections and [1]). More specifically, consider the correlation diagram in which the $2n$ Fermions are grouped into two groups of $n$ Fermions such that each group is Laughlin intra-correlated and there are no inter-correlations between the Fermions of the two groups. The edge-matrix of this correlation diagram (or multi-graph) is

$$E_n := \begin{bmatrix} 2D_n & 0 \\ 0 & 2D_n \end{bmatrix}.$$

Henceforth, we call this the *Moore-Read configuration*. For example, in the case of $N = 8$, the Moore-Read configuration (corresponding to $E_4$) is:

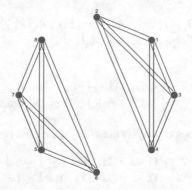

The Moore-Read state, $N = 8$.

© Springer Nature Switzerland AG 2018
S. Mulay et al., *Strong Fermion Interactions in Fractional Quantum Hall States*, Springer Series in Solid-State Sciences 193,
https://doi.org/10.1007/978-3-030-00494-1

The correlation function associated with our Moore-Read configuration is the homogeneous polynomial $G_n := Symm_N(\mu_n)$, where

$$\mu_n := \delta(z, E_n) = \prod_{1 \leq i < j \leq n} (z_i - z_j)^2 \prod_{n+1 \leq k < \ell \leq 2n} (z_k - z_\ell)^2.$$

Since all the pair-correlations in this configuration are of even potency, Theorem 7 assure that the above $G_n$ is indeed a nonzero homogeneous polynomial. More interestingly, by the results of [2], the associated (Quinn) trial wave function $\Psi_Q := G_n \cdot \prod_{1 \leq i < j \leq N} (z_i - z_j)$ is a numerical multiple of the trial wave function $\Psi_{MR}$ (and hence equivalent to $\Psi_{MR}$). Here, we have opted to denote the correlation polynomial as $G_n$ rather than the notation $G_Q$ used for it in the first chapter as well as in [1]. For the benefit of our readers, we present an elementary proof of this equivalence. In what follows, $n$ is tacitly assumed to be a positive integer, $z_1, z_2, \ldots, z_{2n}$ are assumed to be indeterminates and $z$ stands for $(z_1, \ldots, z_{2n})$. We begin by recalling some definitions.

**Definitions**

1. $\det(M)$ denotes the determinant of a square matrix $M$.
2. Given a $(2n) \times (2n)$ skew-symmetric matrix $[a_{ij}]$, the *Pfaffian* of $[a_{ij}]$ is

$$Pf(a_{ij}) = \frac{1}{2^n n!} \sum_{\sigma \in S_{2n}} sgn(\sigma) \prod_{i=1}^{n} a_{\sigma(2i-1)\sigma(2i)},$$

   where $sgn(\sigma)$ is the sign of a permutation $\sigma$ (see, e.g., [3]).
3. Let $z_{ij} := z_i - z_j$ for $1 \leq i, j \leq 2n$. Let $Pf(z_{ij}^{-1})$ denote the Pfaffian of the $(2n) \times (2n)$ skew-symmetric matrix $[a_{ij}]$, where $a_{ii} = 0$ for $1 \leq i \leq 2n$ and

$$a_{ij} := z_{ij}^{-1} = \frac{1}{z_i - z_j} \quad \text{for } 1 \leq i \neq j \leq 2n.$$

4. For $A \subseteq [2n] = \{1, 2, \ldots, 2n\}$, let $A^c := [2n] \setminus A$. Let $\mathcal{T}_n$ denote the collection of all subsets of $[2n]$ of size $n$ that contain 1.
5. For $\sigma \in S_{2n}$, define

$$A(\sigma) := \begin{cases} \{\sigma(i) \mid 1 \leq i \leq n\} & \text{if } \sigma^{-1}(1) \leq n, \\ \{\sigma(i) \mid n+1 \leq i \leq 2n\} & \text{otherwise.} \end{cases}$$

6. Let $\mathcal{M}_{2n}$ denote the set of *perfect matchings*, i.e., the permutations $\sigma \in S_{2n}$ such that

$$\sigma(2i - 1) < \sigma(2i) \quad \text{for } 1 \leq i \leq n \text{ and}$$
$$\sigma(2i - 1) < \sigma(2i + 1) \quad \text{for } 1 \leq i \leq n - 1.$$

*Remark* For $1 \leq i < j \leq 2n$, let $x_{ij}$ be an indeterminate and let $B$ denote the $(2n) \times (2n)$ skew-symmetric matrix whose $(i, j)$-th entry is $x_{ij}$ for $1 \leq i < j \leq 2n$

and whose $(i, i)$-th entry is 0 for $1 \le i \le 2n$. The determinant of $B$ is a polynomial in the indeterminates $x_{ij}$ having coefficients in $\mathbb{Z}$ and as such, it is well-known to be a square. The Pfaffian of $B$ is that square-root of the determinant of $B$ in which the coefficient of the power-product $x_{12}x_{34}\cdots x_{(2n-1)(2n)}$ is a positive integer. This description being universal, it easily specializes to any skew-symmetric matrix of size $2n$.

**Lemma 1** *Let the notation be as above. Then the following holds.*

(i) *We have*

$$Pf(z_{ij}^{-1}) = \sum_{\sigma \in \mathcal{M}_{2n}} \frac{sgn(\sigma)}{\prod_{i=1}^{n}(z_{\sigma(2i-1)} - z_{\sigma(2i)})}.$$

(ii) *symgrp$(E_n)$ is the subgroup of $S_{2n}$ consisting of the permutations $\sigma \in S_{2n}$ such that either $\sigma([n]) = [n]$ or $\{\sigma(i) \mid n+1 \le i \le 2n\} = [n]$.*

(iii) *We have*

$$G_n = |symgrp(E_n)| \cdot \sum_{A \in \mathcal{T}_n} \left( \prod_{i<j \in A}(z_i - z_j)^2 \prod_{k<\ell \in A^c}(z_k - z_\ell)^2 \right).$$

(iv) *The trial wave functions $\Psi_Q$ and $\Psi_{MR}$ are equivalent if and only if*

$$\sum_{A \in \mathcal{T}_n} \left( \prod_{i<j \in A}(z_i - z_j)^2 \prod_{k<\ell \in A^c}(z_k - z_\ell)^2 \right) = \Upsilon Pf(z_{ij}^{-1}) \prod_{1 \le i < j \le 2n}(z_i - z_j)$$

*for some nonzero rational number $\Upsilon$.*

*Proof* Assertion (i) readily follows from a well-known alternative formulation of the Pfaffian (see, e.g., http://en.wikipedia.org/wiki/Pfaffian, 2001). Recall that $[n]$ stands for the set $\{1, \ldots, n\}$. Let $\sigma \in S_{2n}$. Observe that $(z_{\sigma(1)} - z_{\sigma(j)})$ divides $\delta(z, E_n)$ in $\mathbb{Z}[z_1, \ldots, z_{2n}]$ for $2 \le j \le n$, if and only if either $\sigma([n]) \subseteq [n]$ or $\sigma([n]) \subseteq \{n+1, \ldots, 2n\}$. Therefore, $\sigma \in symgrp(E_n)$ if and only if $\sigma([n]) \subseteq [n]$ or $\sigma([n]) \subseteq \{n+1, \ldots, 2n\}$. Thus assertion (ii) holds. For $\sigma \in S_{2n}$, we clearly have $A(\sigma) \in \mathcal{T}_n$ and

$$\sigma(\mu_n) = \prod_{i<j \in A(\sigma)}(z_i - z_j)^2 \prod_{k<\ell \in A(\sigma)^c}(z_k - z_\ell)^2.$$

For $\alpha, \beta \in S_{2n}$, it is straightforward to verify that $A(\alpha) = A(\beta)$ if and only if $\beta = \alpha\theta$ for some $\theta \in symgrp(E_n)$. By first summing the $\sigma(\mu_n)$ as $\sigma$ ranges over a fixed left-coset of $symgrp(E_n)$ in $S_{2n}$ and then summing over all these left-cosets, the equality asserted in (iii) follows. Assertion (iv) readily follows from assertion (iii). □

*Remarks* It is left to the reader to verify that $|symgrp(E_n)| = 2(n!)^2$ (this particular fact will not be used in what follows). The equality displayed in (iv) of Lemma 1 is established in [4] using various conformal field theories. Here, as in [2], we provide a self-contained proof using only elementary observations about determinants and elementary combinatorial properties of permutations. Before proceeding to the proof, we revisit the simplest nontrivial example that was presented in the seventh section of the first chapter.

*Example* Recall the case of $n = 2$, i.e., $N = 4$. Here, we have

$$G_{MR} := z_{12}z_{13}z_{24}z_{34} - z_{12}z_{14}z_{23}z_{34} + z_{13}z_{14}z_{23}z_{24}.$$

As is easily verified, $\mathcal{T}_n = \{A_1 := \{1, 2\}, A_2 := \{1, 4\}, A_3 := \{1, 3\}\}$. Also, note that $symgrp(E_2)$ is the set

$$\{id, (1, 2), (3, 4), (1, 2)(3, 4), (1, 3)(2, 4), (1, 4)(2, 3), (1, 3, 2, 4), (1, 4, 2, 3)\},$$

where $id$ is the identity permutation and the other members are expressed as products of disjoint cycles. The full set of representatives for the left-cosets of $symgrp(E_2)$ in $S_4$ is: $\{\tau_1 := (1, 2), \ \tau_2 := (1, 3), \ \tau_3 := (1, 4)\}$. Clearly, we have

$$\tau_r(\delta(z, E_2)) = \prod_{i < j \in A_r} (z_i - z_j)^2 \prod_{k < \ell \in A_r^c} (z_k - z_\ell)^2$$

for $1 \leq r \leq 3$. Hence $G_2 = 8 \cdot (z_{12}^2 z_{34}^2 + z_{13}^2 z_{24}^2 + z_{14}^2 z_{23}^2)$. By a direct expansion, it can be easily verified that $G_2 = 16 \cdot G_{MR}$.

**Definitions** Let $z_1, \ldots, z_{2n}$ and $T$ be indeterminates.

1. For $1 \leq i \leq n$, define $x_i := z_{2i-1}$ and $y_i := z_{2i}$. Let $X = \{x_1, \ldots, x_n\}$ and $Y = \{y_1, \ldots, y_n\}$.
2. Let $L(X, Y) := [a_{ij}]$ be an $n \times n$ matrix with

$$a_{ij} := (x_i - y_j)^{-1} \text{ for } 1 \leq i \leq n \text{ and } 1 \leq j \leq n.$$

3. Define polynomials $F(X)$ and $F(Y)$ by

$$F(X) := \prod_{1 \leq i < j \leq n} (x_i - x_j) \quad \text{and} \quad F(Y) := \prod_{1 \leq i < j \leq n} (y_i - y_j),$$

   and then define $F(X, Y) := F(X)F(Y)$.
4. Let $f(T) := (T - y_1) \cdots (T - y_n)$ and let $f_T$ denote the $T$-derivative of $f(T)$.

**Lemma 2** *With the above notation, the following holds.*

*(i) $X, Y$ are disjoint ordered sets of indeterminates and*

$$\prod_{1 \leq i < j \leq 2n} (z_i - z_j) = (-1)^{\frac{n(n-1)}{2}} F(X, Y) f(x_1) \cdot f(x_2) \cdots f(x_n).$$

*(ii) We have*

$$\left( \prod_{1 \leq i < j \leq 2n} (z_i - z_j) \right) \cdot \det(L(X, Y)) = F(X, Y)^2 = F(X)^2 F(Y)^2.$$

*Proof* We prove assertion (i) by induction on $n$. If $n = 1$, then $F(X, Y) = 1$ and hence (i) trivially holds. Assuming $n \geq 2$, our assertion readily follows from the induction hypothesis and the observation:

$$(z_{2n-1} - z_{2n}) \prod_{i=1}^{n-1} (z_{2i} - z_{2n-1}) = (x_n - y_n) \prod_{i=1}^{n-1} (y_i - x_n) = (-1)^{n-1} f(x_n).$$

Next, we prove (ii). Let $L := L(X, Y)$. Multiplying the $i$-th row of $L$ by $f(x_i)$ for all $1 \leq i \leq n$, we obtain the $n \times n$ matrix

$$M := [b_{ij}], \quad \text{where} \quad b_{ij} := \frac{f(x_i)}{(x_i - y_j)}.$$

Since each $b_{ij}$ is homogeneous of degree $n - 1$ in $x_1, \ldots, x_n, y_1, \ldots, y_n$, the determinant of $M$ is a homogeneous polynomial of degree $n(n - 1)$. It is straightforward to verify that

$$\det(M) = f(x_1) \cdot f(x_2) \cdots f(x_n) \cdot \det(L).$$

It is now evident that if we put $x_i = x_j$ (resp. $y_i = y_j$) for any $i < j$, then $\det(M)$ vanishes. By the unique factorization property of polynomials in $\mathbb{Z}[X, Y]$, each prime polynomial $x_i - x_j$ (resp. $y_i - y_j$) with $i < j$ must divide $\det(M)$. For distinct pairs $(i, j)$ with $i \neq j$, the corresponding $x_i - x_j$ (resp. $y_i - y_j$) are mutually relatively prime non-associate prime polynomials and hence $F(X, Y)$ must divide $\det(M)$ in $\mathbb{Z}[X, Y]$. Since $F(X, Y)$ is also homogeneous of degree $n(n - 1)$ in $x_1, \ldots, x_n, y_1, \ldots, y_n$, we infer that $\det(M) = c F(X, Y)$ for some integer $c$. Substituting $x_i = y_i$ for all $1 \leq i \leq n$ in each entry of $M$ reduces $M$ to a diagonal matrix with $f_T(y_j)$ as its $(j, j)$-th entry and hence $\det(M)$ evaluates to the product $f_T(y_1) f_T(y_2) \cdots f_T(y_n)$. On the other hand, substituting $x_i = y_i$ for all $1 \leq i \leq n$ in $F(X, Y)$ yields $F(Y)^2$. Consequently, $f_T(y_1) f_T(y_2) \cdots f_T(y_n) = c F(Y)^2$. From this last equality, it readily follows that $c = (-1)^e$, where $e = n(n - 1)/2$, and hence (ii) follows from (i). $\qquad \square$

*Remark* Assertion (ii) of the above lemma can also be derived from a determinantal identity found in [5]; this derivation is demonstrated in [2].

**Definitions** Consider $Z := \{z_1, \ldots, z_{2n}\}$ as an unordered set.

1. Given a partition $X, Y$ of the set $Z$ into sets $X$ and $Y$ of cardinality $n$ each, fix a listing $\sigma$ of the elements of $X$ to obtain the ordered set $X_\sigma := \{x_1, \ldots, x_n\}$ and fix a listing $\tau$ of the elements of $Y$ to obtain the ordered set $Y_\tau := \{y_1, \ldots, y_n\}$.

Define the corresponding ordering $(\sigma, \tau)$ of $Z$ by declaring $Z_{(\sigma,\tau)} = \{z_1, \ldots, z_{2n}\}$, where

$$z_{2i-1} := x_i \quad \text{and} \quad z_{2i} := y_i \quad \text{for} \ \ 1 \leq i \leq n,$$

and then define

$$\Delta(X_\sigma, Y_\tau) := \prod_{1 \leq i < j \leq 2n} (z_i - z_j).$$

2. Let $\mathcal{P}_2$ be the set of all unordered partitions $\{X, Y\}$ of $Z$ into two parts of cardinality $n$ each.

**Lemma 3** *Let $Z$, $(\sigma, \tau)$, etc., be as in the above definition.*

(i) *The product $\Delta(X_\sigma, Y_\tau) \cdot \det(L(X_\sigma, Y_\tau))$ is independent of $(\sigma, \tau)$, i.e.,*

$$\Delta(X_\sigma, Y_\tau) \cdot \det(L(X_\sigma, Y_\tau)) = \Delta(X, Y) \cdot \det(L(X, Y)).$$

(ii) *The product $\Delta(X, Y) \cdot \det(L(X, Y))$ is symmetric in $X$ and $Y$, i.e.,*

$$\Delta(X, Y) \cdot \det(L(X, Y)) = \Delta(Y, X) \cdot \det(L(Y, X)).$$

(iii) *We have*

$$\sum_{A \in \mathcal{T}_n} \Delta(A, A^c) \det(L(A, A^c)) = \sum_{A \in \mathcal{T}_n} \left( \prod_{i<j\in A} (z_i - z_j)^2 \prod_{k<\ell\in A^c} (z_k - z_\ell)^2 \right).$$

*Proof* Note that $F(X, Y)^2 = F(X)^2 F(Y)^2$, with $F(X)^2$ and $F(Y)^2$ invariant under any ordering of $X$ and $Y$, respectively. Hence (i) follows from (ii) of the above Lemma 2. Likewise, since it is clear that $F(X, Y) = F(Y, X)$, assertion (ii) also follows from (ii) of the above Lemma 2. Finally, summing the equalities in (ii) of the above Lemma 2 corresponding to all possible members of $\mathcal{P}_2$ leads to the equation:

$$\sum_{\{X,Y\}\in\mathcal{P}_2} \Delta(X, Y) \det(L(X, Y)) = \sum_{\{X,Y\}\in\mathcal{P}_2} F(X)^2 F(Y)^2.$$

Given $\{X, Y\}$, there is a unique $A \in \mathcal{T}_n$ such that $\{X, Y\} = \{A, A^c\}$. In view of (ii), assertion (iii) now readily follows. $\qquad\qquad\square$

**Definitions** We continue to use the above notation. Let $A \in \mathcal{T}_n$ and $\sigma \in S_n$. Enumerate $A$ so that $A := \{u_1, \ldots, u_n\}$ with $u_1 < u_2 < \cdots < u_n$ and enumerate $A^c$ so that $A^c := \{v_1, \ldots, v_n\}$ with $v_1 < v_2 < \cdots < v_n$.

1. Define

$$b(A, \sigma) := \left\{ \min\{u_i, v_{\sigma(i)}\} \, \middle| \, 1 \leq i \leq n \right\},$$
$$c(A, \sigma) := \left\{ \max\{u_i, v_{\sigma(i)}\} \, \middle| \, 1 \leq i \leq n \right\}.$$

2. Suppose $b(A, \sigma) = \{b_1 < \cdots < b_n\}$ and $c(A, \sigma) = \{c_1 < \cdots < c_n\}$. Given $i \in [n]$, let $j \in [n]$ be the unique integer (depending on $i$) such that $b_i = \min\{u_j, v_{\sigma(j)}\}$ and let $\tau(i) \in [n]$ be the unique integer such that $c_{\tau(i)} = \max\{u_j, v_{\sigma(j)}\}$. For $1 \leq i \leq n$, let $\tau(A, \sigma)(i) := \tau(i)$.

3. Define the permutation $\Theta_{(A,\sigma)} \in S_{2n}$ by

$$\Theta_{(A,\sigma)}(i) := \begin{cases} u_j & \text{if } i = 2j - 1 \text{ with } 1 \leq j \leq n, \\ v_{\sigma(j)} & \text{if } i = 2j \text{ with } 1 \leq j \leq n. \end{cases}$$

By $\Theta_A$, we mean $\Theta_{(A,id)}$, where $id$ denotes the identity permutation.

*Example* Let $n = 3$, $A := \{1, 4, 6\}$ and let $\sigma$ be the transposition $(2, 3)$. Then, $A^c = \{2, 3, 5\}$, $b(A, \sigma) = \{1, 3, 4\}$ and $c(A, \sigma) = \{2, 5, 6\}$. Observe that $\Theta_A$ is the transposition-product $(3, 4)(5, 6)$ and $\Theta_{(A,\sigma)}$ is the 4-cycle $(3, 4, 5, 6)$. Observe that $\tau(A, \sigma)$ is the transposition $(2, 3)$.

**Lemma 4** *With the above notation, the following holds.*

(i) *For $A \in \mathcal{T}_n$,*

$$\Delta(A, A^c) = sgn\,(\Theta_A)\, F(z), \quad where \quad F(z) := \prod_{1 \leq i < j \leq 2n} (z_i - z_j).$$

(ii) *For $A \in \mathcal{T}_n$ and $\sigma \in S_n$,*

$$sgn\,\big(\Theta_{(A,\sigma)}\big) = sgn\,(\Theta_A)\, sgn(\sigma).$$

(iii) *Letting $B := b(A, \sigma)$, $C := c(A, \sigma)$ and $\tau := \tau(A, \sigma)$, we have $B \in \mathcal{T}_n$, $C = [2n] \setminus B$ and $\tau \in S_n$. Moreover, $\Theta_{(B,\tau)} \in \mathcal{M}_{2n}$.*

*Proof* Fix a pair $(A, \sigma) \in \mathcal{T}_n \times S_n$. From the definitions of $\Delta(A, A^c)$ and $\Theta_A$, it is apparent that $\Delta(A, A^c) = \Theta_A\,(F(z))$. Since $\theta(F(z)) = sgn(\theta) \cdot F(z)$ for all $\theta \in S_{2n}$, assertion (i) follows. Let $\alpha \in S_{2n}$ be defined by

$$\alpha(i) := \begin{cases} i & \text{if } i = 2j - 1 \text{ with } 1 \leq j \leq n, \\ 2\sigma(j) & \text{if } i = 2j \text{ with } 1 \leq j \leq n. \end{cases}$$

Examining the disjoint cycle decompositions of $\alpha$ and $\sigma$, it follows at once that $sgn(\alpha) = sgn(\sigma)$. Since $\Theta_{(A,\sigma)} = \Theta_A \circ \alpha$, assertion (ii) follows. Lastly let $B, C$ be as in (iii). Since $1 \in A$, we must have $1 = u_1 = \min\{u_1, v_{\sigma(1)}\}$ and hence $b_1 = 1$. Additionally, since

$$\{u_i, v_{\sigma(i)}\} \cap \{u_j, v_{\sigma(j)}\} = \emptyset \quad \text{for } 1 \leq i < j \leq n,$$

we have $|B| = n = |C|$ and $B \cap C = \emptyset$. So, $B \in \mathcal{T}_n$ and $C = [2n] \setminus B$. Clearly $\tau$ is a permutation of the set $[n]$, i.e., $\tau \in S_n$. The definition of $\tau(A, \sigma)$ ensures that $b_i < c_{\tau(i)}$ for $1 \leq i \leq n$. Consequently, $\Theta_{(B,\tau)}$ is a member of the set $\mathcal{M}_{2n}$. $\square$

**Definition** Assertion (iii) of the above lemma allows us to define a function

$$\phi : \mathcal{T}_n \times S_n \rightarrow \mathcal{M}_{2n}$$

which maps an ordered pair $(A, \sigma)$ to the permutation $\phi(A, \sigma)$ defined by

$$\phi(A, \sigma) := \Theta_{(b(A,\sigma),\, \tau(A,\sigma))}.$$

*Examples* Let $n = 3$ and $A := \{1, 4, 6\}$.

1. Let $\sigma$ be the transposition $(2, 3)$. As noted in the example just before Lemma 4, $b(A, \sigma) = \{1, 3, 4\}$ and $\Theta_{(A,\sigma)}$ is the 4-cycle $(3, 4, 5, 6)$. Observe that $\tau(A, \sigma)$ is the transposition $(2, 3)$ and $\phi(A, \sigma)$ is the 3-cycle $(4, 6, 5)$.
2. Let $\sigma$ be the 3-cycle $(1, 3, 2)$. Then, $b(A, \sigma) = \{1, 2, 3\}$. It is straightforward to verify that $\Theta_{(A,\sigma)}$ is the 5-cycle $(2, 5, 6, 3, 4)$ and $\tau(A, \sigma)$ is the transposition $(1, 2)$. Consequently, $\phi(A, \sigma)$ is the 3-cycle $(2, 5, 3)$.

**Lemma 5** *Let the notation be as above.*

(i) *$\phi(A, \sigma) = \phi(b(A, \sigma), \tau(A, \sigma))$ for all $(A, \sigma) \in \mathcal{T}_n \times S_n$.*
(ii) *For each $\rho \in \mathcal{M}_{2n}$, we have $|\phi^{-1}(\rho)| = 2^{n-1}$.*
(iii) *Suppose $(A, \sigma) \in \mathcal{T}_n \times S_n$. Let $\rho := \phi(A, \sigma)$ and $\theta := \Theta_{(A,\sigma)}$. Then*

$$\frac{sgn(\theta)}{\prod_{i=1}^{n}(z_{\theta(2i-1)} - z_{\theta(2i)})} = \frac{sgn(\rho)}{\prod_{i=1}^{n}(z_{\rho(2i-1)} - z_{\rho(2i)})}.$$

*Proof* Assertion (i) is an easy consequence of the definition of $\phi$. Fix a permutation $\rho \in \mathcal{M}_{2n}$. Let $(A, \sigma)$, $b(A, \sigma)$, $c(A, \sigma)$ be as in the above definition of $\phi$. Observe that $\phi(A, \sigma) = \rho$ if and only if for each $i \in [n]$, there is a $i_j \in [n]$ with

$$\min\{u_{i_j}, v_{\sigma(i_j)}\} = b_i = \rho(2i - 1) < \rho(2i) = \max\{u_{i_j}, v_{\sigma(i_j)}\}.$$

So, the pairs $(A, \sigma)$ in $\phi^{-1}(\rho)$ are characterized by the property that $u_1 = 1$ and

$$\left\{\{u_r, v_{\sigma(r)}\} \mid r \in [n]\right\} = \left\{\{\rho(2s - 1), \rho(2s)\} \mid s \in [n]\right\}.$$

Thus $\phi^{-1}(\rho)$ has exactly $2^{n-1}$ members as asserted in (ii). Next, let the notation be as in (iii). Let $\eta := \rho^{-1}\theta$. As noted in the proof of (ii), we have

$$\left\{\{\theta(2r - 1), \theta(2r)\} \mid r \in [n]\right\} = \left\{\{\rho(2s - 1), \rho(2s)\} \mid s \in [n]\right\}$$

and hence there is a unique permutation $\alpha \in S_n$ such that

$$\{\eta(2r - 1), \eta(2r)\} = \{2\alpha(r) - 1, 2\alpha(r)\} \quad \text{for } 1 \leq r \leq n.$$

From the basic property of *sgn*, the equality asserted in (iii) is clearly equivalent to

$$(*) \qquad \frac{sgn(\eta)}{\prod_{i=1}^{n}(z_{\eta(2i-1)} - z_{\eta(2i)})} = \frac{1}{\prod_{i=1}^{n}(z_{2i-1} - z_{2i})}.$$

Let $I := \{r \in [n] \mid \eta(2r - 1) = 2\alpha(r)\}$. Let $\tau$ denote the product of mutually disjoint transpositions $(2\alpha(r) - 1, \ 2\alpha(r))$ as $r$ ranges over $I$ ($\tau$ is the identity permutation if $I$ is empty) and let $\pi := \tau\eta$. Then,

$$\pi(i) = \begin{cases} 2\alpha(r) - 1 & \text{if } i = 2r - 1 \text{ with } r \in [n], \\ 2\alpha(r) & \text{if } i = 2r \text{ with } r \in [n]. \end{cases}$$

Observe that if $(a_1, \ldots, a_s)$ is a cycle of length $s$ appearing in the disjoint cycle decomposition of $\alpha$, then each of $(2a_1 - 1, \ldots, 2a_s - 1)$ and $(2a_1, \ldots, 2a_s)$ appears in the disjoint cycle decomposition of $\pi$. From this observation, it follows at once that $\pi$ is an even permutation, i.e., $sgn(\pi) = 1$. Consequently, $sgn(\eta) = sgn(\tau)sgn(\pi) = (-1)^{|I|}$. Since we also have

$$\prod_{i=1}^{n}(z_{\eta(2i-1)} - z_{\eta(2i)}) = (-1)^{|I|} \cdot \prod_{i=1}^{n}(z_{2i-1} - z_{2i}),$$

the equality $(*)$ stands verified. Thus assertion (iii) holds. $\qquad\qquad\square$

**Theorem** *Let the notation be as above. Then, we have*

$$\sum_{A \in \mathcal{T}_n} \left( \prod_{i<j \in A} (z_i - z_j)^2 \prod_{k<\ell \in A^c} (z_k - z_\ell)^2 \right) = 2^{n-1} \cdot Pf(z_{ij}^{-1}) \prod_{1 \le i < j \le 2n} (z_i - z_j).$$

*In particular, the trial wave functions* $\Psi_Q$ *and* $\Psi_{MR}$ *are equivalent.*

*Proof* Thanks to assertion (iv) of Lemma 1, equivalence of $\Psi_Q$ and $\Psi_{MR}$ follows from the above asserted equality. By (iii) of Lemma 3,

$$\sum_{A \in \mathcal{T}_n} \Delta(A, A^c) \det(L(A, A^c)) = \sum_{A \in \mathcal{T}_n} \left( \prod_{i<j \in A} (z_i - z_j)^2 \prod_{k<\ell \in A^c} (z_k - z_\ell)^2 \right).$$

Using (i) and (ii) of Lemma 4 and using the definition of $\det(L(A, A^c))$, we express the sum appearing on the left of the above equation as

$$(**) \qquad \left[ \prod_{1 \le i < j \le 2n} (z_i - z_j) \right] \cdot \sum_{(A,\sigma) \in \mathcal{T}_n \times S_n} \frac{sgn\left(\Theta_{(A,\sigma)}\right)}{\prod_{i=1}^{n}(z_{\Theta_{(A,\sigma)}(2i-1)} - z_{\Theta_{(A,\sigma)}(2i)})}.$$

As a consequence of assertions (i), (ii) and (iii) of Lemma 5, $(**)$ is equal to

$$\left[ \prod_{1 \le i < j \le 2n} (z_i - z_j) \right] \cdot 2^{n-1} \cdot \sum_{\rho \in \mathcal{M}_{2n}} \frac{sgn(\rho)}{\prod_{i=1}^{n} (z_{\rho(2i-1)} - z_{\rho(2i)})}.$$

Now in view of (i) of Lemma 1, the equality

$$2^{n-1} \cdot Pf(z_{ij}^{-1}) \cdot \left[ \prod_{1 \le i < j \le 2n} (z_i - z_j) \right] = \sum_{A \in \mathcal{T}_n} \left( \prod_{i < j \in A} (z_i - z_j)^2 \prod_{k < \ell \in A^c} (z_k - z_\ell)^2 \right)$$

stands verified. Thus our assertion is established.                                                    □

*Remarks*

1. The above theorem in conjunction with the remarks following Lemma 1 show
   that

   $$G_n = 2^n \cdot (n!)^2 \cdot Pf(z_{ij}^{-1}) \prod_{1 \le i < j \le 2n} (z_i - z_j)$$

   and hence

   $$\Psi_Q = 2^n \cdot (n!)^2 \cdot Pf(z_{ij}^{-1}) \prod_{1 \le i < j \le 2n} (z_i - z_j)^2.$$

2. For $n = 3$, it is interesting to note that $\Psi_Q$ is also the trial wave function of a
   non-minimal configuration of 6 Fermions (see Sect. 2.4.1) in the Jain IQL state
   with filling factor $\nu = 3/7$.

# Appendix B
# Questions

Below, we list some currently unresolved problems that naturally arise in dealing with the correlation diagrams and their associated correlation functions. These problems are purely mathematical in nature. Our search of the existing literature on the closely related mathematics seems to suggest that these problems have not been investigated, or for that matter, even posed. Of course, our search far from being exhaustive and our expertise limited, some of our questions possibly have known answers. In the following list, we use the notation and definitions introduced in the second chapter. It is tacitly assumed that $N$, $d$ are integers such that $N \geq 3$, $d \geq 3$ and $z$ stands for indeterminates $z_1, \ldots, z_N$.

1. Let $V := (d_1, \ldots, d_N)$, where $d_i$ is a nonnegative integer for $1 \leq i \leq N$. What requirements on $(N, V)$ are necessary and sufficient for $E(N, V)$ to be nonempty? More ambitiously, what is the cardinality of $E(N, V)$?

2. What are the integers $m$ for which the coefficient of $X^m$ in the polynomial

$$\mathfrak{G}(N, d, X) := \frac{(1 - X^{N+1})(1 - X^{N+2}) \cdots (1 - X^{N+d})}{(1 - X^2)(1 - X^3) \cdots (1 - X^d)}$$

   is nonzero? In other words, determine the support of $\mathfrak{G}(N, d, X)$. Of course, it is even more useful to obtain a meaningful lower bound for these coefficients.

3. For which pairs $(E_1, E_2) \in E(N, \leq d) \times E(N, \leq d)$ does there exist a nonzero rational number $\Upsilon$ such that

$$Symm_N(\delta(z, E_1)) = \Upsilon \cdot Symm_N(\delta(z, E_2))?$$

4. Here is an open problem related to (i) of Theorem 7 (see Sect. 2 of Chap. 2). Under what conditions on $E \in E(N)$ does there exist an $a := (a_1, \ldots, a_N)$ in $\mathbb{R}^N$ such that $\delta(\sigma(a), E)$ is positive for all $\sigma \in S_N$?

5. In the direction opposite to (i) of Theorem 7, we may ask: for what $E \in E(N)$, if any, is $Symm_N(\delta(z, E))$ a sum of squares of real polynomials? For a graph-theoretic investigation of this problem, the reader is referred to [6].

© Springer Nature Switzerland AG 2018
S. Mulay et al., *Strong Fermion Interactions in Fractional Quantum Hall States*, Springer Series in Solid-State Sciences 193,
https://doi.org/10.1007/978-3-030-00494-1

6. Suppose $\mathfrak{m} : m_1 \leq \cdots \leq m_q$, $\varepsilon$ are as in the definition of $\mathfrak{m}$-excellence and assume that $\varepsilon$ satisfies the $\mathfrak{m}$-excellence property (1). Then, it is of great interest to formulate a property of $\varepsilon$ that simultaneously generalizes the $\mathfrak{m}$-excellence properties (2), (3) in such a way that if $\varepsilon$ satisfies this particular property, then the symmetrization of $\mu(z, \varepsilon)$ as well as that of $\mu(z, \varepsilon)^{-1}$ is assured to be nonzero.

7. In the context of Theorem 10, we pose the following question. Let $m$, $n$ be positive integers such that $m \leq N - 2$. What are the $m \times n$ matrices $A := [a(i, j)]$ with nonnegative integer entries $a(i, j)$ such that letting

$$
E := \begin{bmatrix} 0 & A \\ A^T & 0 \end{bmatrix},
$$

$Symm_N (\delta(z, -E))$ (note the negative sign!) is nonzero?

8. In the context of Theorem 12, consider the following question: for given $N$, $d$, what are the necessary and sufficient requirements on $\lambda$ which ensure the existence of $\mathfrak{m} : m_1 \leq \cdots \leq m_q$ and an $\mathfrak{m}$-excellent function $\varepsilon$ such that the corresponding $\mu(z, \varepsilon)$ has total degree $\lambda$ as well as the $z_i$-degree at most $d$ for each $i$? Obviously, a complete answer to this question will render Theorems 8–9 even more useful.

9. Here are some questions that arise in the context of Theorem 13. Fix an integer $\lambda$ and define

$$
J(N, d, \lambda) := \{E \in E(N, \leq d) \mid \|E\| = 2\lambda \quad \text{and} \quad Symm_N (\delta(z, E)) \neq 0\}.
$$

(i) What is the minimum of $\{ bound(E) \mid E \in J(N, d, \lambda)\}$?

(ii) For what $\lambda$ is there an $E \in J(N, d, \lambda)$ with $bound(E) \leq 2$?

(iii) What is the minimum of $\{ [S_N : symgrp(E)] \mid E \in J(N, d, \lambda)\}$?

(iv) What is the maximum of the set

$$
\{ [Symgrp(E) : symgrp(E)] \mid E \in J(N, d, \lambda)\}?
$$

(v) For what $\lambda$ is there an $E \in J(N, d, \lambda)$ such that $|suppt(E)| \leq 3$?

10. Construction of correlation functions associated with configurations containing quasielectrons (see the last section of Chap. 2) poses the following problem. Given integer $m$ with $3 \leq m \leq 1 + (N/2)$ and a half-integer $L \in \Lambda(N, m)$, what restrictions on the triple $(N, m, L)$ are necessary and sufficient for there to exist a matrix $A \in E(N, \leq 2(N - 1) - m)$ with $\|A\| = Nd - 2L$ such that $A$ has $2D_{N-m}$ as a diagonal block and $Symm_N (\delta(z, A))$ is nonzero?

# Appendix C
# Computations

Below, we list some computational procedures that we have used. It must be pointed out that none of the authors claim any expertise in computation. So, these algorithms are most likely not very efficient or economical; yet, they can be of help to the reader in understanding how the explicit correlation polynomials presented as examples in the second chapter are computed. In what follows, by a 'procedure', we mean a MAPLE procedure.

**1. Computation of Correlation Functions**.

Let $E := [a_{ij}]$ be an $N \times N$ symmetric matrix with nonnegative integer entries and each of whose diagonal entries is 0. In the procedure below, $E$ is represented as a list of $N - 1$ lists:

$$E := [[a_{12}, \ldots, a_{1N}], \ldots, [a_{i(i+1)}, \ldots, a_{iN}], \ldots, [a_{(N-1)N}]].$$

The function $G$ of the procedure is the polynomial $Symm_N(\delta(z, E))$ defined in the second chapter. The function $wyG$ of the procedure computes the expression of $G$ as a polynomial in $y_1, \ldots, y_{N-1}$ (as in the second chapter).

```
with(combinat);
v := proc (N::nonnegint)
     options operator, arrow, function_assign;
     [seq(z[i], i = 1 .. N)]
          end proc:
h := proc (R::nonnegint, N::nonnegint, E::list, L::list)
     options operator, arrow, function_assign;
     product((L[R]-L[R+j])^E[R][j], j = 1 .. N-R)
          end proc:
g := proc (E::list, N::nonnegint, L::list)
     options operator, arrow, function_assign;
     product(h(R, N, E, L), R = 1 .. N-1)
          end proc:
```

© Springer Nature Switzerland AG 2018
S. Mulay et al., *Strong Fermion Interactions in Fractional Quantum Hall States*, Springer Series in Solid-State Sciences 193,
https://doi.org/10.1007/978-3-030-00494-1

```
G := proc (E::list)
     options operator, arrow, function_assign;
     simplify(sum( expand(
        g(E, nops(E)+1, permute(v(nops(E)+1))[i])),
        i = 1 .. nops(permute(v(nops(E)+1))))))
        end proc:
ele := proc (E::list)
        options operator, arrow, function_assign;
        {convert(G(E, nops(E)+1), 'elsymfun')}
           end proc:
eq2 := proc (N::nonnegint)
        options operator, arrow, function_assign;
        expand(product(x+z[j], j = 1 .. N))
           end proc:
eq1 := proc (N::nonnegint)
        options operator, arrow, function_assign;
        expand((x+t/N)^N+sum(y[k]*(x+t/N)^(N-k-1), k = 1 .. N-1))
           end proc:
eq := proc (E::list)
       options operator, arrow, function_assign;
       [seq(coeff(eq2(nops(E)+1), x, i) =
           coeff(eq1(nops(E)+1), x, i), i = 0 .. nops(E))]
           end proc:
u := proc (E::list, p::nonnegint) option remember;
     if p = 1 then algsubs(eq(E)[1], ele(E))
        else subs(eq(E)[p], u(E, p-1))
        end if end proc:
wyG := proc (E::list) options operator, arrow, function_assign;
       simplify(u(E, nops(E)+1))
           end proc;
```

The trial wave function $\psi$ for the system of $N$ Fermions corresponding to $E$ is computed by the procedure below. Note that the correlation polynomial $G$ of the configuration is one of the inputs required by $\psi$.

```
with(LinearAlgebra):
fermi := proc (N)
        options operator, arrow, function_assign:
              Determinant(VandermondeMatrix(z, N))
              end proc:
psi := proc (G, N)
       options operator, arrow, function_assign;
           (-1)^(N*(N-1)/2)*fermi(N)*G
              end proc;
```

## 2. Correlation Functions of Minimal Configurations.

Let $m$ and $n$ be positive integers and $N := mn$. In the fourth section of the second chapter, the correlation function of a minimal configuration of $N$ Fermions in the $\nu = n/(2pn \pm 1)$ IQL state, is defined to be a polynomial of the form $Symm_N(\delta(z, E))$, where $E = M(m, n, a, c)$. Of course, the above procedure can be used to compute this correlation polynomial; but the procedure presented below is more economical in computational terms because it exploits the symmetries of these minimal configurations. Thus, with appropriate inputs $m, n, a, c$, the function $G$ outputs the corresponding correlation polynomial (which is equivalent to $Symm_N(\delta(z, E))$; see Theorem 13). The associated trial wave function can then be computed using the above procedure $\psi$.

```
with(GroupTheory):
g1 := proc (m::posint, n::posint)
        options operator, arrow, function_assign;
            {[seq([i*m+1, i*m+2], i = 0 .. n-1)]}
            end proc:
g2 := proc (m::posint, n::posint)
        options operator, arrow, function_assign;
            {[seq([seq(i*m+j, j = 1 .. m)], i = 0 .. n-1)]}
            end proc:
g3 := proc (m::posint, n::posint)
        options operator, arrow, function_assign;
            {[seq([i, m+i], i = 1 .. m)]}
            end proc:
g4 := proc (m::posint, n::posint)
        options operator, arrow, function_assign;
            {[seq([seq(i*m+j, i = 0 .. n-1)], j = 1 .. m)]}
            end proc:
g := proc (m::posint, n::posint)
        options operator, arrow, function_assign;
            'union'(g1(m, n), g2(m, n), g3(m, n), g4(m, n))
            end proc:
gro := proc (m::posint, n::posint)
        options operator, arrow, function_assign;
            Group(g(m, n), supergroup = SymmetricGroup(m*n))
            end proc:
coreps := proc (m::posint, n::posint)
            options operator, arrow, function_assign;
                map(Representative,
                RightCosets(gro(m, n), SymmetricGroup(m*n)))
                end proc:
with(combinat):
v := proc (m::posint, n::posint, s::posint)
        options operator, arrow, function_assign;
            [seq(z[coreps(m, n)[s][r]], r = 1 .. m*n)]
            end proc:
h := proc (k::posint, m::posint, n::posint, E::list, L::list)
        options operator, arrow, function_assign;
            mul((L[k]-L[k+j])^E[k, k+j], j = 1 .. m*n-k)
            end proc:
```

```
gg := proc (E::list, m::posint, n::posint, L::list)
     options operator, arrow, function_assign;
          mul(h(k, m, n, E, L), k = 1 .. m*n-1)
          end proc:
G0 := proc (E::list, m::posint, n::posint)
     options operator, arrow, function_assign;
          simplify(sum(expand(
          gg(E, m, n, v(m, n, t))), t = 1 .. nops(coreps(m, n))))
          end proc:
with(LinearAlgebra):
eta := proc (u, v)
     options operator, arrow;
          abs(signum(u-v))
          end proc:
dee := proc (r::posint, s::posint)
     options operator, arrow, function_assign;
          Matrix(r, s, eta)
          end proc:
B := proc (r::posint, s::posint, m::posint, a, c)
     options operator, arrow, function_assign;
     (2*(1-abs(signum(r-s)))*a+abs(signum(r-s))*c)*dee(m, m, eta)
          end proc:
bl := proc (m::posint, n::posint, a, c)
     options operator, arrow, function_assign;
          [seq([seq(B(i, j, m, a, c), j = 1 .. n)], i = 1 .. n)]
          end proc:
M := proc (m::posint, n::posint, a, c)
     options operator, arrow, function_assign;
          Matrix(bl(m, n, a, c))
          end proc;
G := proc (m::posint, n::posint, a, c)
     options operator, arrow, function_assign;
          G0(M(m, n, a, c), m, n)
          end proc;
```

## 3. Total Angular Momenta for Systems with QE.

For systems of $N$ Fermions with $m$ QE in $\nu = 1/3$ IQL, it is necessary to compute the allowed total angular momenta of the system. The following procedure is based on the description presented in the fifth section of the second chapter; it outputs the set $\Lambda(N, m)$ of the values of allowed total angular momenta.

```
with(QDifferenceEquations):
with(powseries):
F := proc (l, N, q)
     options operator, arrow, function_assign;
          expand(expand(QBinomial(2*l+1, N, q^2)))
          end proc;
f := proc (l, N, q)
     options operator, arrow, function_assign;
          quo(F(l, N, q), q^(N*(2*l-N+1)), q)
          end proc;
```

```
g := proc (l, N, q)
    options operator, arrow, function_assign;
        expand((q^2-1)*f(l, N, q)/q^2)
        end proc;
h := proc (l, N)
    options operator, arrow, function_assign;
    [seq([(1/2)*i, coeff(g(l, N, q), q,i)], i = 0 .. N*(2*l-N+1))]
        end proc;
lambda := proc (l, N)
    local t, s; option remember;
        t := NULL; for s to nops(h(l, N)) do if h(l, N)[s][2] <> 0
        then t := {h(l, N)[s][1]} union t end if end do; return t
        end proc;
Lambda := proc (N, m)
        options operator, arrow, function_assign;
            lambda((1/2)*N+1/2-(1/2)*m, m)
            end proc;
```

# References

1. S.B. Mulay, J.J. Quinn, M.A. Shattuck, Correlation diagrams: an intuitive approach to correlations in quantum Hall systems. J. Phys. Conf. Ser. **702**, 1–10 (2016)
2. S.B. Mulay, J.J. Quinn, M.A. Shattuck, Λ generalized polynomial identity arising from quantum mechanics. Appl. Appl. Math. **11**, 576–584 (2016)
3. M. Hazewinkel, *Pfaffian, Encyclopedia of Mathematics* (Springer, Berlin, 2012). ISBN 978-1-55608-010-4
4. A. Cappelli, L.S. Georgiev, I.T. Todorov, Parafermion Hall states from coset projections of abelian conformal theories. Nucl. Phys. B **599**(3), 499–530 (2001)
5. C. Krattenthaler, Advanced determinant calculus, Sém. Lothar. Combin. (The Andrews Festschrift) **42** (1999). Article B42q
6. P. Alexandersson, B. Shapiro, Discriminants, symmetrized graph monomials, and sums of squares. Exp. Math. **21**, 353–361 (2012)

# Index

**A**
Admissible, 74
Anharmonic, 11

**B**
Balanced configuration, 31
Bound, 95

**C**
Cayley, 52
Chen, 3, 5, 17
Chern–Simons flux, 4
Clebsch, 20
Configuration, 27
Correlation, 2, 4, 13–19, 21–23
Correlation-factor, 14
Correlation function, 27
Coulomb, 1, 2, 4, 5, 9, 13, 18

**D**
Dirac, 2
Discriminant, 49, 56
Dixmier, 56

**E**
Eckart, 14
Equivalent configurations, 28
Excellent, 62
Exchange-energy, 1

**F**
Fermi, 1, 2, 13, 19

**G**
Gordon, 20
Graph-monomial, 27

**H**
Haldane, 2, 4, 6–8, 13, 17
Harmonic, 9
Hartree, 2
Hermite reciprocity, 53

**I**
Invariant, 46

**J**
Jain, 2–10, 12, 14, 16–18, 21

**L**
Landau, 2–5, 13–15, 17
Laughlin, 2–4, 6, 9, 11–14, 18, 19, 21, 22
Laughlin configuration, 28
Lorentz, 16

**M**
Magneto-resistivity, 3
Minimal configuration, 32
Moore, 11, 12, 15, 21
Multi-graph, 27
Multiplet, 8, 10, 11, 13, 19, 21, 112

© Springer Nature Switzerland AG 2018
S. Mulay et al., *Strong Fermion Interactions in Fractional Quantum
Hall States*, Springer Series in Solid-State Sciences 193,
https://doi.org/10.1007/978-3-030-00494-1

**O**
Order, 56

**P**
Pauli, 1, 12
Pfaffian, 15, 138
Psuedopotential, 8–11, 13, 23

**Q**
Quasielectron, 2, 12, 13, 18
Quasihole, 2, 12
Quasiparticle, 17

**R**
Read, 11, 12, 15, 21
Regular multi-graph, 29

**S**
Self-energy, 1, 2
Semi-invariant, 41
Silin, 2
Sitko, 6, 8
Skew invariant, 92
Sommerfeld, 1
Spherical geometry, 13
State, 111

**T**
Tsui, 2

**W**
Weighted degree, 40, 41
Weighted homogeneous, 41
Wigner, 14

Printed in the United States
By Bookmasters